Crane Hazards and Their Prevention

David V. MacCollum, P.E., C.S.P.

American Society of Safety Engineers
1800 East Oakton Street
Des Plaines, Illinois 60018

Managing Editor: Michael F. Burditt
Project Editor: Jeri Ann Stucka
Text and Graphics Composition: Sue Knopf, Graffolio
Cover: Michael Burditt and Sue Trebswether, ASSE

Library of Congress Cataloging-in-Publication Data

MacCollum, David V., 1923–
    Crane hazards and their prevention/David V. MacCollum.
        p.   cm.
    Includes bibliographical references (p.   ) and index.
    ISBN 0-939874-95-4
    1. Cranes, derricks, etc.––Safety measures.   I. Title.
TJ1363.M275   1993
621.8'7'0289––dc20   93-39816
                                    CIP

The text of this book is composed in Baskerville, with display type set in Helvetica Compressed.

Cover photographs by Michael Burditt

Printed in the United States on acid-free, recycled paper. ♲
First Edition
Acknowledgments for reprint permissions follow diagrams when appropriate.

    11 10 09 08 07 06 05   1 2 3 4 5 6 7 8

*This book is dedicated to my loving wife of fifty years,
Nancy, who has always edited my writings.*

Illustrations

Figure 3-1   Telescoping Boom Mobile Hydraulic Crane [13]

Figure 3-2   Truck-Mounted Hydraulic Boom Crane [14]

Figure 3-3   Truck-Mounted Latticework Boom Crane [14]

Figure 3-4   Flatbed-Truck Mounted Pedestal Hydraulic Boom [15]

Figure 3-5   Truck-Mounted Remote Control Articulated Boom [15]

Figure 3-6   Flatbed-Mounted Trolley Boom [15]

Figure 3-7A   Overhead Track-Mounted Cranes (Gantry, Overhead Traveling, Semi-Gantry, Wall, Cantilever Gantry) [16]

Figure 3-7B   Overhead Track-Mounted Cranes (Tower, Portal, Hammerhead) [17]

Figure 3-8   Examples of Monorails and Underhung Hoists [17]

Figure 3-9   Wheel-Mounted Straddle Lift [18]

Figure 3-10   Hammerhead Tower Crane [18]

Figure 3-11   Derrick Crane [19]

Figure 3-12   Digger Derrick [19]

Figure 3-13   Examples of Aerial Platforms [20]

Figure 4-1A
4-1B
4-1C   Table 232-1 of the *National Electrical Safety Code,* ANSI C2, American National Standards Institute, Inc. [27-29]

Figure 4-2   Request #IR 159 dated April 11, 1974, and its Interpretation in *National Electrical Safety Code Interpretations, 1961-1977 Inclusive,* NESC Interpretations 1961-1977, American National Standards Institute, Inc. [29]

Figure 4-3A
    4-3B
    4-3C   Mapping the Danger Zone [30-31]

Figure 4-4   Insulated Link   [33]

Figure 5-1   Upset Crane [39]

Figure 5-2   Interlocks Limit Boom Movement [43]

Figure 5-3   Aerial Basket Upset with Outriggers Retracted [44]

Figure 6-1   Anti-Two-Blocking Device [52]

Figure 7-1   Crushed Cab [54]

Figure 7-2   Control Levers for Two-Wheel/Four-Wheel Steer [55]

Figure 7-3   No Emergency Quick Release Exit [57]

Figure 8-1   Unsafe Access [60]

Figure 9-1   Close Clearance Between Crane Cab and Truck Carrier [66]

Figure 9-2   Warning Label for Shear Point [66]

Figure 9-3   U.S. Patent 2,892,549, Safety Guards for Cranes and the Like [67]

Figure 9-4   U.S. Patent 3,812,978, Crane Safety Barrier [68]

Figure 9-5   Design Concept to Prevent Pinch Point Injuries [69]

Figure 10-1   Typical Mobile Hydraulic Crane Cab Control Station [73]

Figure 10-2   Remote Control [75]

Figure 10-3   Remote Control with Small Levers [75]

Figure 10-4   Control Accessibility [75]

Figure 10-5   Unguarded Controls [77]

Figure 10-6   Guarded Controls [77]

Figure 11-1   Travel Alarms and Wheel Guards for Straddle Cranes [84]

Figure 13-1   Unsafe Hook/Safe Hook [96]

Figure 13-2A   Hook that Allows Load Loss [96]

Figure 13-2B   Hook that Prevents Load Loss [96]

Figure 14-1 Unsafe Boom Disassembly [108]

Figure 14-2 Boom Disassembly Label for Crane Cab [109]

Figure 14-3 Boom Disassembly Label for Crane Boom [110]

Figure 21-1 Oil Well Servicing Derrick [134]

Figure 21-2 Recommended Anchor Locations [134]

Figure 21-3 Anchor Capacity Requirements for Each Zone [135]

Figure 21-4 Anchor Location Diagram [135]

Figure 21-5 Catenary Method [136]

Figure 25-1A
  25-1B Hand Signals [149-150]

Foreword

This book is particularly well-timed since the rapidly evolving requirements of world trade now present new challenges in terms of both imports and exports. Competitive product challenges and new legal requirements for safety, such as the EC Directives, require world class products capable of use in many environments by those from many cultural backgrounds. Such challenges provide new opportunities if the design and operational objectives can be clarified.

I have always been impressed by the design and capability of cranes and related equipment whether serving industrial or construction site needs. However, performance capability achievements have not always been matched by the safety performance of cranes. There are well-known reasons why there have been some deficiencies in the overall design safety aspects of cranes and the resultant creation of occupational safety problems in field use, service, maintenance, and repair of cranes. Past problems should not imply criticism, unless the problems continue to be manifested in the future. Past problems, constructively reviewed, serve as lessons that facilitate corrective and preventive action. Identification of failure modes is merely a first step in identifying remedies.

Common crane hazards that have resulted in injuries, damage, and lawsuits can serve as a key safety checklist in the design review of new cranes and associated equipment. Elimination of inherent design hazards and a significant reduction in the risk of operating cranes would facilitate the institution of reasonable and effective occupational safety programs for cranes. Thus, loss control is a major objective of this publication.

The author of this book is an outspoken advocate for safety, based on some fifty years of full-time professional level experience. He has specifically and graphically identified more than forty hazards and remedies with which he has had detailed personal involvement. This is not a book on theory or management. It deals with specific and practical problems that have become manifest in the field and which deserve closer attention. Thus, the book is a realistic approach to loss control and risk management. The author is to be complimented for writing this book as a form of quick learning and experience retention that could be effectively used by professional safety specialists. It is in a form that could evoke the kind of critical comment and discussion that is necessary for the professional development of an increasingly complex field of endeavor. The book and its contents are organized and succinctly presented in a manner well-suited to those who can make a difference in terms of product improvement and occupational safety. In essence, the safety objectives are so clarified that the old problems need not reappear in the future in world class products.

GEORGE A. PETERS

Peters and Peters
Attorney, Counselors at Law
Santa Monica, CA

Preface

This hands-on book gives a practical look at specific hazards associated with cranes and their use that have taken many lives and have resulted in many tragic disabilities. Over fifty hazards are discussed and a sampling of safety requirements and other references are listed for each hazard. Examples of hazard prevention measures that my experience has shown to be effective in controlling these hazards are also listed. New ways to control hazards are continually being tried and are proven effective. The hazard prevention measures in this book are suggested to aid and stimulate others in using their own personal ingenuity and creativity in applying available technology to come up with other needed hazard prevention measures in the hope that much of the pain and suffering now experienced can be ended.

The writing of this book has been prompted by the need to share my over fifty years' experience in analyzing hazards. Cranes play a very vital role in construction productivity, but when management and supervisory staff are not aware of inherent hazards in design and/or use and have not had sufficient guidance to overcome them, their use can become a disaster.

This book is being published not only for safety professionals, but to assist designers, manufacturers, crane rental firms, management involved in the use of cranes, crane operators, property owners on whose premises cranes are used, and electric utility companies whose powerlines are often struck by cranes. It is not intended to be a performance or compliance guide but has been written as an informational source for some of the available hazard prevention measures and design alternatives. It is hoped that it will be a ready reference for those who seek continuous improvement in reducing and controlling crane hazards, as it shows what others are doing and what can be done to make crane operations safer, more productive, and avoid injuries that involve unwanted publicity, government inquiry, fines, and often litigation.

In today's world, governmental or standards-making bodies cannot write regulations and standards fast enough to cover all hazardous circumstances as they arise. Technology moves too fast. We need improved communication of hazard and injury information from all available sources, so those who must assure a safe workplace are made aware of what others in similar circumstances are experiencing in terms of injury and property damage, and, most importantly, the most effective, currently available options to prevent reoccurrence of injury from the same hazard.

Table of Contents

	Illustrations	**iv**
	Foreword	**vii**
	Preface	**ix**
Chapter 1	**Crane "Accident" Causes**	**1**
	Reasons for Injury, Death, or Property Damage [1]	
Chapter 2	**Hazard Prevention Concepts**	**7**
Chapter 3	**Crane Types**	**13**
Chapter 4	**Powerline Contact**	**21**
	Powerline Contact by Hoist or Boom [21]	
	Powerline Contact by Aerial Lifts [32]	
Chapter 5	**Upset**	**35**
	Upset Due to Overloading [35]	
	Failure to Use Outriggers / Soft Ground and Structural Failure [40]	
	Lift Simulation [43]	
	Aerial Lift Upset When Outriggers Not Extended [44]	
Chapter 6	**Two-Blocking**	**47**
Chapter 7	**Operator's Station**	**53**
	No Crush-Resistant Cabs [53]	
	No ROPs on Rough-Terrain Cranes [54]	
	No Operator Seat Belt [56]	
	Emergency Exit System [57]	
	Flying or Swinging Objects [58]	
Chapter 8	**Access**	**59**
	Unsafe Access to Cab, Housing, and Cab Roof [59]	
	Unsafe Walkways on Overhead Bridge Cranes [61]	
Chapter 9	**Pinch Points and Nip Points**	**63**
	Pinch Points [63]	
	Unguarded Nip Points [69]	

Chapter 10   Controls   71

Control Confusion/Inadvertent Activation of Controls [71]
Controls on a Conductive Tether or Controls Reachable from Ground [74]
Unintended Movement of Aerial Lifts [76]

Chapter 11   Vision   79

Blind Lifts [79]
Backing [80]
Vision Compromise on Large Track-Mounted Cranes [81]
Vision Compromise on Straddle Cranes and Crew Cab Failure [83]

Chapter 12   Cable and Sheave Damage   85

Sheave-Caused Cable Damage [85]
Cable Kinking [87]
Hoist Cable Binding in Boom Hoist Tip Sheave on Hydraulic Cranes [88]
Breaking of Boom Hoist Cable Causing Latticework Boom to Fall [89]

Chapter 13   Load Loss   91

"Killer" Hooks [91]
Free Fall [97]
Load Loss on Straddle Cranes [99]

Chapter 14   Boom Failures and Disassembly   101

Boom Buckling [101]
Loss of Stowed Jib Booms on Cantilevered Hydraulic Booms [102]
Latticework Boom Toppling Over Backwards [103]
Side Pull [105]
Improper Boom Disassembly on Latticework Boom Cranes [106]

Chapter 15   Transportation   111

Bouncing Booms on Cranes Being Transported [111]
*Adjusting A-Frames on Latticework-Boom Cranes to Accommodate
     Low Underpasses* [112]
Absence of Communication Between Carrier Cab and Crane Cab [112]
Loading and Unloading Cranes From Trailers and Trucks [113]

Chapter 16   Counterweights   115

*Dropping of Removable Counterweights and Extendible Counterweight
     Pinch Point* [115]
Upset of Counterweight-Stabilized Aerial Lifts [116]

Chapter 17   Turntable Undocking and Boom Hinge Pin Assembly Failure   119

Chapter 18   Tower Cranes   121

Upset of Self-Raising Tower Cranes [121]
Footing Failure on Tower Cranes [122]

Chapter 19   Overhead Bridge Cranes   125

Exposed Electrical Trolleys on Overhead Bridge Cranes [125]
Absence of Lockout Systems on Overhead Bridge Cranes [126]

Chapter 20 **Monorails and Underhung Crane Stops** **129**

Chapter 21 **Guyline Anchor Systems** **131**

Chapter 22 **Wind and Weather** **137**

Chapter 23 **Maintenance and Renovation** **139**

Chapter 24 **Personnel Requirements** **143**
Competency of Operators [143]
Training [145]
Competency of Riggers, Signalers, and Others [145]

Chapter 25 **General Hazard Prevention Procedures** **147**

Appendix **153**

Bibliography **157**

Glossary of Definitions **163**

Index **169**

1

Crane "Accident" Causes

Reasons for Injury, Death, or Property Damage

In our highly mechanized world, cranes are the workhorses that have increased productivity and economic growth in construction, mining, logging, maritime operations, and maintenance of production and service facilities. In large metropolitan areas it is not unusual to see several crane booms outlined against the skyline within a few blocks of one other, and in rural areas to see cranes being used in a great variety of ways. Because cranes are so indispensable in today's workplace and because their population increases daily, the prevention of crane hazards is very important. Statistics and data from the technical library of the Hazard Information Foundation, Inc. (HIFI), U.S. Department of Labor, the Occupational Safety and Health Administration (OSHA), the Mine Safety and Health Administration (MSHA), and personal injury litigation were used collectively to show that a crane can become a very dangerous piece of equipment because of inherent hazards that can be encountered during normal working circumstances. Using this data, I have found over fifty hazards that have caused most crippling injuries and deaths and have targeted them for discussion.

Serious injury or death that occurs repeatedly from similar circumstances should be considered epidemic. These occurrences should be examined to identify hazards so appropriate hazard prevention measures can be initiated in the same diligent manner that the medical profession examines a disease or infection to develop a vaccine or antibiotic for its prevention or control. Unfortunately, at the work site, many occurrences are often labeled as "freak accidents" because operating personnel are sometimes totally unaware of similar repetitive occurrences because of their wide dispersal. This wide geographical dispersal over a long time period and the collection of hazard/injury data by so many agencies have both contributed to a widespread lack of awareness of hazards associated with crane use that can cause unintentional injury, death or property damage. The use of the word "accident" has been avoided as it implies that it was unforeseeable, unavoidable and, therefore, not preventable. In much scientific work, "accident" is gradually being replaced with "unintentional injury, death or property damage." This change was mentioned in a footnote in *The Injury Fact Book* by Baker, O'Neil, Ginsburg and Li, 2nd Edition (Oxford University Press, 1992), p. 36.

Some cranes are inadvertently and unintentionally designed with built-in hazards. Although our design engineers are very competent when it comes to structural, mechanical, or electrical aspects of crane design, they do not always have sufficient training in the most complex component of design: the people who are going to use the crane. People make errors in judgment when working with cranes because cranes are so versatile and can be used in so many different ways. Sometimes, too much reliance has been placed upon the ability of those using cranes to make last-minute adjustments to avoid injury or death. It has been found that operators cannot always do this. A well-known human factors specialist, Alphonse Chapanis, has changed the old saying "To err is human, to forgive, divine" to "To err is human, to forgive, *design*." A key principle in preventing unintentional injury, death, or property damage is to modify design to reduce the hazard. This is often more feasible than attempting to alter the behavior of individuals to recognize and avoid a hazard. When safety is included in design, the errors of those who use the equipment can be overcome by the same technology space engineers use when designing spacecraft. (See bibliographic listings 101-107 for references on human factors.)

The safety of crane operators and those working with cranes is as important as that of our astronauts. If someone makes the wrong move or hits the wrong switch, there is no reason, with today's technology, for that person's life to be put in jeopardy. It is not an insurmountable task to make the workplace safe if the crane's limitations and the limitations of those using them are considered, especially when planning is done before work commences. My personal investigation and review of several thousand injuries and deaths involving cranes over a period of over fifty years has shown that there is often little planning for crane use in a particular set of work circumstances. Another factor is that, in many circumstances, the responsibility for crane safety is delegated to the lowest operating level, where hazard prevention options are very limited. The worker may not perceive the hazard or may lack the authority to take steps to eliminate the hazard. (See Chapter 18, "Crane Safety on Construction Job Site," *Construction Safety Management and Engineering*, American Society of Safety Engineers, Des Plaines, IL, Darryl C. Hill, editor. These two factors in themselves are nothing more than an invitation to disaster. (See bibliographic listings 108-119 for references on pre-job planning.)

Because the man/machine aspect of de-

sign has not received the attention it should have had, some cranes arrive at the workplace with inherent hazards, and it becomes a challenge to identify these hazards and apply appropriate hazard prevention measures for their elimination.

The actual number of cranes in use in the United States is estimated to be somewhere between 150,000 and 300,000, and worldwide, close to 600,000. When a crane made for specific uses that has a population of, say, only 1,000 — a small percentage of the total crane population — is found to have caused repeated death or serious injury, its reliability should be considered unacceptable. The public was outraged when unsafe gas tanks on the Pinto automobiles caused 125 deaths. Considering the total population of Pintos, this casualty rate was .004 percent of the total population of 3,000,000 Pintos. A crane with a population of 1,000 that has caused 10 deaths or injuries has a rate of one percent, and nobody seems to be outraged.

Another reason for the lack of awareness of the extent of crane-related injury and property damage is that only a small segment of our society is directly involved in the use of cranes. The work force using cranes is largely an unseen minority. Construction, mining, lumbering, and maritime industries employ less than fifteen percent of the work force but account for roughly sixty percent of workplace deaths and injuries. Less than five percent of our nation's work force is employed in construction, but this five percent accounts for twenty percent of all workplace fatalities. Almost every family in this country knows of someone who participated in Desert Storm in the Persian Gulf where, as a result of smart weapons and battle planning, we sustained phenomenally few casualties for combat. With safe crane design and job safety planning before all crane use, crane hazards can be targeted as effectively as the military did its combat hazards.

By my best estimate, from examining existing government and National Safety Council injury data and litigation involving death or permanently disabling injury, crane hazards are the source of about one-fourth to one-third of casualties in construction and maintenance activities.

The statistics for injury occurrences relied upon in this book are taken from the not-for-profit organization the Hazard Information Foundation, Inc. injury and fatality data that has been litigated. These comprehensive statistics include unreported, self-employed incidents. Also relied upon are statistics from the U.S. Department of Labor Occupational Safety and Health Administration (OSHA) and the Mine Safety and Health Administration (MSHA) injury and death data for the last forty years. Since most states are under state plans, similar data from the National Safety Council and the National Institute for Safety and Health (NIOSH) creates a matrix to identify occurrence averages, which are roughly listed here:

- 25% powerline contacts
- 19% crane upset from various causes
- 12% two-blocking
- 10% types of boom failures
- 7% loss of control of load from winch failure
- 3% hook latch failures
- 24% various absence of guarding, unsafe access, blind zones and other design defects.

The identification of many of the hazards listed in this book has been aided by examining data of some 2,000 personal injury liability lawsuits involving cranes that have been filed over the last 34 years, of which I am aware. This database gives a current profile of the types of hazards in crane-related injuries for 34 years.

- 1970 through 1979 — 18%
- 1980 through 1989 — 30%
- 1990 through 1999 — 40%
- 2000 through 2004 — 12%

Personal injury litigation involving cranes is a costly fact-of-life and must be discussed and dealt with in any discussion of crane hazards. The immediate response to liability litigation is often reactive rather than proactive — "fire the referee" or "kill the messenger of bad news." A very valuable untapped source of hazard information available to responsible management and employers who want to make continuous improvement in products and workplaces is the safety research done for courtroom presentation to find how many similar occurrences there have been, why the injured and others were unaware of the danger, and what hazard prevention measures would have prevented the injury.

In today's world, the only way to avoid litigation is to eliminate hazardous circumstances that can cause injury, as the legal barriers that have shielded those who have created hazardous circumstances are steadily being overturned by the courts. As society begins to recognize that we cannot afford to contaminate our environment, it is also coming to realize that our workforce is not expendable. These opinions are influencing personal injury litigation.

At the end of most chapters I have listed some representative cases to show how the justice system is focusing on hazards that cause injury. From the standpoint of hazard prevention, the outcome of the litigation is unimportant. What is important is that it is another source of hazard information.

Many of the old excuses used in the courtroom for not applying available hazard prevention measures are beginning to be found invalid:

1. The excuse that the rest of the industry is not using available hazard prevention measures may not be acceptable.

2. The excuse that the safety device was not totally and absolutely reliable and was not used for that reason may not be acceptable if the device provides reasonable protection most of the time. You would not discontinue the use of pneumatic tires because they blow out or puncture, and you would not discard electronic navigation equipment in aircraft because it sometimes momentarily fails. It is better that you have reasonable protection most of the time than no protection at all.

3. The excuse that the purchaser chose to delete the safeguard may not be acceptable if the manufacturer/seller knew the safeguard would overcome a life-threatening hazard.

4. The excuse that there was no written policy requiring the use of specific hazard prevention measures to overcome a life-threatening hazard may not be acceptable.

5. The excuse that the employer cannot be sued under workmen's compensation law may not be acceptable if the employer fails to provide necessary

safeguards for hazards that were identified by the employer's safety staff. This has been found to be fraudulent concealment of known workplace hazards.

6. The excuse that the employer cannot be sued under workmen's compensation law may not be acceptable if a supervisor was made aware of a life-threatening hazard prior to injury and did not allow the workers to initiate hazard prevention measures they recommended.

7. The excuse by a landowner and owner of a crane that the contractor was using the crane in an unsafe manner may not be acceptable if the landowner retains control over the contractor's performance.

8. The excuse by a contractor that the injured was working for another contractor may not acceptable if the first contractor created the hazardous circumstance.

9. The excuse that the employer cannot be sued under workmen's compensation law may not be acceptable if the conduct of the employer is so outrageous that it amounts to an intentional act.

10. The excuse that records concerning a hazard are unavailable may not be acceptable if it is found that they were intentionally destroyed.

Our world as we know it today, particularly in activities with high injury, death, or property damage, is in a constant state of change because of almost instant communication. When disaster strikes, the whole world knows about it and knows what needs to be done to avoid future catastrophic occurrence. Saying that the hazard was not controlled because no regulations or standards existed covering the hazard is not good enough. Because governmental and voluntary standards-making bodies are unable to write rules, regulations, and standards fast enough to provide reasonable guidance, we must look to our own research to provide the answers. We must develop the ability to gather and transfer hazard information from all sources to management and the operating level without waiting for standards and regulations to be formulated. Unfortunately, much hazard information is sometimes concealed for years by protective orders that lock away in darkness the actual dangers that people in the workplace face and the available hazard prevention measures prescribed to overcome them. We must use all sources, including litigation, when gathering information on crane hazards and see that there is a free exchange of this information among designers, manufacturers, sellers, rental companies, owners, employers who use cranes, and operators. Only then will we see a reduction in unintentional injury, death, and property damage involving cranes.

The hazards are listed in two groups, those associated with most cranes and those associated with specific types of cranes.

Hazards Associated with Most Cranes

Backing
Blind lifts
Boom buckling
Boom overloads from loads not freely
    suspended
Cable kinking
Control confusion/inadvertent
    activation of controls or lack of a
    control
Dropping of removable counter-
    weights and extendible counter-
    weight pinch point
Failure to use outriggers/soft ground
    and structural failure
Flying or swinging objects
Loose bolts
No crush-resistant cabs
No emergency exit
No operator seat belt
Pinch point
Powerline contact
Sheave-caused cable damage
Side pull
Turntable failure or undocking
Two-blocking
Unguarded nip points
Unsafe access to cab, housing, and
    cab roof
Unsafe hooks
Upset due to overloading

Hazards on Specific Types of Cranes

Absence of communication between
    carrier cab and crane cab

Absence of lockout systems on overhead bridge cranes

Adjusting A-frames on latticework-boom cranes to accommodate low underpasses

Aerial lift upset when outriggers not extended

Bouncing booms on cranes being transported

Breaking of boom-hoist cable causing latticework boom to fall

Controls on a conductive tether or controls reachable from the ground

Exposed electrical trolleys on overhead bridge cranes

Footing failure on tower cranes

Free fall

Hoist cable binding in boom hoist tip sheave on hydraulic cranes

Improper boom disassembly on latticework boom cranes

Inadequate stops on monorails and underhung cranes

Latticework boom toppling over backwards

Load loss on straddle cranes

Loading and unloading from trailers and trucks

Loss of stowed jib booms on cantilevered hydraulic booms

No ROPS on rough-terrain cranes

No cab swing pin

No leveling device

Powerline contact by aerial work platforms

Unintended movement of aerial lifts

Unsafe walkways on overhead bridge cranes

Upset of counterweight-stabilized aerial lifts

Upset of self-raising tower cranes

Vision compromise on large track-mounted cranes

Vision compromise on straddle cranes

Wind blowing against long raised booms and some tower cranes

People have to know about and understand hazards to overcome them. Each of the above crane hazards can be controlled by pre-job safety planning, innovative application of OSHA requirements or other appropriate safety standards or measures, safety training, and innovative safe design improvements, or a combination of all of these.

General OSHA requirements for cranes and derricks are as follows:

General Industry: Sections 1910.179 through 1910.181
Maritime Operations: Sections 1917.45 through 1917.46
Longshoring: Section 1918.74
Construction: Section 1926.550

This book does not attempt to address rigging or procedures for handling loads but is directed towards the discussion of specific hazards inherent to cranes in general and a few hazards that apply only to a specific type of crane. There are many references that discuss safe rigging, and most operator's manuals tell specifically how the crane has to be set up for safe use. The following topics are discussed for each hazard:

Definition

Description

**Risks Presented by Hazard
Available Hazard Prevention
Measures**

These measures may or may not be required by OSHA or used by other organizations. They are listed to provide information to those who choose to voluntarily adopt hazard prevention measures to optimize their own safety performance. They are only a sampling of what can be done. New methods and technology are constantly being developed. It is hoped that the information provided will spark individual creativity to develop measures that fit specific needs.

OSHA Requirements

After a hazard is identified, look closely at OSHA's rules and regulations. Most of the time you will find that they cover the hazard in question and give good guidance for its control.

ANSI Requirements

Other References
Military standards, patents, etc. are good sources for available safety technology.

Suggested Design Criteria
Design improvements have already been incorporated in some makes or models or are available from aftermarket suppliers. Review of product specifications gives good guidance for picking the right crane for a specific use. This book is not intended to recommend or qualify one crane model or make-over the design of another make or model.

Representative Litigation
These cases are listed to show that this hazard did cause death or injury that eventually led to litigation. It is hoped that the listing of these cases will develop an awareness that the greatest deterrent to litigation is adequate hazard prevention. These descriptions are my personal understanding and interpretation of the hazard that led to injury. The listing of these cases identifies hazards and does not detail the legal aspects of the cases as I am not a lawyer. If readers want to know more about the legal aspects of the case, it is suggested they obtain the advise and counsel of a licensed lawyer in the appropriate jurisdiction.

Neither the author nor publisher is listing this information as a source of legal guidance. The intent is only to show the relationship of a hazard or hazardous circumstances to legal action.

Hazard Prevention Concepts

A hazard is an unsafe physical condition that exists in normal crane usage. If a hazard has not been eliminated by the designer, it can be controlled at the workplace if it is understood that a hazard exists and lies in wait in one of three modes:

1. *Dormant:* A dormant hazard is an undetected hazard created either by design or crane use.

2. *Armed:* An armed hazard is a dormant hazard that has not been detected during the planning phase and becomes armed and ready to cause harm when the crane is used in certain work circumstances as it presents a source of energy capable of causing injury.

3. *Active:* An active hazard is an armed hazard triggered into action by the right combination of factors and is releasing energy in a manner to cause harm. At this point it is too late to implement safeguards to avoid or prevent death or injury or for escape.

System safety is an approach to hazard prevention that involves the detection of: failure modes of equipment in the environment in which it may be used; and, the variable of human performance that may arise during its use. Its objective is to improve design so it will be user-friendly and not error-provocative and vulnerable to the environment. (For references on system safety see bibliographic listings 125-133.) A convenient, hands-on approach to system safety hazards is to realize that they fall into seven basic categories. These need to be anticipated by both design and construction managers with the expertise of a safety professional. These categories are:

- Natural: Environmental and physical aspects of the world we live in and must cope with
- Mechanical
- Electrical
- Chemical
- Radiant
- Biological
- Artificial Intelligence: Use of computer software programs to automate crane movements within safe limits.

By examining each of these facets of exposure to hazards and applying them to the design of the crane, the facility to be constructed, and the construction scheduling, hazard identification and prevention is achievable. The following five basic steps for hazard prevention, based on system safety concepts, are listed in order of decreasing importance:

1. *Design to Eliminate or Minimize the Hazard.* The major effort through the design/planning phase of any project must be to design equipment with appropriate safeguards to render it and its use fail-safe. Redundancy and back-up components or measures are needed. An older crane can be retrofitted with current safety improvements developed

since its manufacture. Design improvement is more effective than attempting to rely on changing human behavior.

2. *Guard the Hazard.* Hazards that cannot be totally eliminated through design or planning must be reduced to an acceptable level of risk by the use of appropriate safety devices to guard, isolate, or otherwise render the hazard effectively inert or inaccessible. "Guarding" covers a broad range of applied technology. For example, workers can be effectively guarded against falls from an elevation by using proper scaffolding. If scaffolding cannot be erected at a given location, the next best choice is often to select a crane-supported, suspended work platform, provided all requirements for the crane, the platform, and procedures set forth by OSHA are met. Designers should consider the hazard and then examine available technology to control the hazard. When design is unsafe, management should examine available after-market technology for hazard prevention. Many crane hazards can be significantly reduced by incorporating available technology at the time of design or by the owner's selection of after-market accessories. There no magic index number or formula for an acceptable level of risk, as it is a value judgment that generally reflects the conscience of management. An excellent reference that provides a comprehensive discussion of acceptable risk is *Of Acceptable Risk, Science and the Determination of Safety,* by William W. Lowrance (William Kaufmann, Inc., Los Altos, California, 1976). Methods of guarding a hazard or the danger zone it creates include:

a. Installing screens or covers over moving parts.

b. Insulating the hazard so unintentional physical contact with energized conductors does not prove injurious or fatal. This form of guarding includes insulated links for cranes and line covers for energized powerlines.

c. Installing fences, guardrails, etc. that restrict entry to a danger zone or a fall from an elevation.

d. Installing interlocks that stop or deactivate the system when overruns are made or when guards or covers are removed. The anti-two-blocking device, which is designed to stop the hoist line, is an application of an interlock. Interlocks have been included in United Underwriters Laboratories' standards since the 1960s to prevent shock hazards. Many applications of interlocks are used to isolate energy sources and dangerous mechanical movement, such as found in washing machines and clothes dryers that stop when the cover is opened.

e. Installing sensors to give an alert when someone intrudes into a danger zone. These include infrared motion detectors, high frequency sound, radar, light barriers, and pressure sensors. A few examples of available sensor technology are security devices that rely upon infrared sensors, bar coding used in supermarkets, temperature controls, and stress gauges.

f. Installing a crush-resistant cab and restraint system that provides a safe sanctuary for the operator in the event of upset by enclosing him/her in a protective frame. The development of rollover protection and crush-resistant design can be extended to crane cab design by tapping already available technology.

g. Isolating dangerous locations where moving parts or hazardous energy sources exist.

h. Installing fail-safe systems such as hydraulic boom snubs that prevent the boom from tipping backwards over the cab. A nearly vertical lattice boom can tip vertically past 90 degrees when a crane is on an uneven surface and its pendant guys are unable to provide support.

i. Providing redundancy by having a number of safeguards that work together to overcome the hazard. In circumstances where the loss of life or serious injury from a specific hazard is present, it is prudent to provide design where several barriers or safeguards must be breached before harm can result.

j. Installing electrical monitoring systems for electrical or mechanical systems.

3. *Give Warning.* When a hazard cannot be eliminated by applying either the first or second methods, an active/intercessory warning device should be installed that detects the hazard and emits a timely, audible and/or visual warning signal. Examples are alarms, horns, and flashing lights. Warning systems must be of the standard variety, so the meaning of the warning is never misunderstood. Some hazard detection systems not only give audible or visual warning but are wired to stop or turn away movement of the crane boom before it reaches a hazardous position.

Signs and labels are passive warnings. They must be very explicit and state what the hazard is, what harm will result, and how to avoid the hazard. The signs for life-threatening hazards should be as pictorial as possible with the word "DANGER" written in white letters on an oval red background with a black border. Warning systems are not substitutes for the first and second hazard prevention methods. Rather, warnings are best used to alert users to be aware of a specific change of circumstances or of a dormant hazard that cannot be totally eliminated or controlled. Warning should also describe why the specified safeguard must be used. Requirements for signs and labels are set forth in OSHA 1910.145, "Specifications for Accident Prevention Signs and Tags"; OSHA 1926.200, Subpart G, "Signs, Signals, and Barricades"; ANSI Z535.1-1991, *Safety Color Code*, Z535.2-1991, *Environmental and Facility Safety Signs*, Z535.3-1991, *Criteria for Safety Symbols*, Z535.4-1991, *Product Safety Signs and Labels* and Z535.5-1991,

Accident Prevention Tags for Temporary Hazards; and the Society of Automotive Engineers' *SAE Recommended Practices*, SAE J115, "Safety Signs."

Mere vocal warning to use the crane safely does not do the job. Crane owners need to assure that crane operators, riggers, and others who work with cranes are provided with safe operating procedures and given training.

4. *Special Procedures and Training.* When a hazard cannot be eliminated or its risk reduced by guarding or use of intercessory warning devices, system safety hazard prevention demands that special operating procedures, training, and audits be employed to guarantee that a viable, continuing regimen is established and maintained to avoid the hazard. Special procedures and training must be based upon the physical characteristics of specific cranes, warnings posted by the manufacturer on the crane or listed in its operator's manual, and injury data. To assume that perfect human performance can be achieved to avoid injury or death is unrealistic. Blaming the victim does not remove the hazard and fails to prevent injury to the next unsuspecting, unwary, or uninformed individual confronted with a similar circumstance.

5. *Personal Protective Equipment.* Use of gloves, hard hats, safety shoes, aprons, goggles, safety glasses, life lines, life jackets, and other protective equipment at all times will also protect users from injury.

A networking of several of these five controls is often necessary to avoid a life-threatening hazard.

A key concept in the prevention of crane hazards is that each party has a clearly defined duty to fulfill whatever is within its own reasonable span of control, totally independent from what other parties may or may not be doing within their spans of control. Far too often one party attempts to excuse its non-performance by pointing to the failure of other parties to initiate hazard prevention measures within their spans of control. In short, the failure of others in fulfilling their hazard prevention responsibilities is no excuse for not doing your own duty.

The manufacturer must fulfill its span of safety control by designing its equipment so that hazards are reduced by elimination and/or guarding techniques. Additionally, operator's manuals and point-of-operation warnings provide manufacturers the opportunity to inform users of hazards they will encounter when using the machine. Warnings should spell out specific safety requirements. Prospective buyers should be made aware of any inherent hazards posed by the normal use of the equipment. Sales brochures and specifications should address each hazard and acquaint prospective buyers with necessary safeguards. Orientation of buyers as to hazards and available safety features to control them is particularly important, as some purchasing agents, since they usually do not use the equipment they buy, have very little understanding in this regard and are unaware that operator aids and safety features can be long-term money-saving items. Buyers' primary goal is to cut costs, but what many buyers do not realize is that an uncontrolled hazard can cost an employer many times the initial purchase price of a safeguard if injury or death occurs.

Not only safety professionals, but also design engineers, salespeople, rental firms, owners, project engineers, construction managers, construction supervisors, electric utility personnel, crane operators, and all those whose work involves the use of cranes and who have some measure of control over a crane during its entire life cycle — beginning with the design concept to the time it is dismantled for salvage — need to be familiar with all hazards and assure that they personally initiate positive action to eliminate all hazards that are within their individual span of control.

I believe system safety is the backbone of hazard prevention as it can detect specific hazards that may have been otherwise overlooked. During World War II, system safety emerged in the design of aircraft and blossomed in the aerospace technology. During the 1970s, forward-thinking individuals applied system safety concepts to a variety of other products to improve quality, utility, and safety. The "smart" weapons used in the Gulf War proved the value of a system safety hazard analysis in evaluating specific hazards.

Change is needed and must be directed towards identifying and controlling hazards. Federal intervention in workplace safety has become gridlocked by rule-making and compliance, creating an adversarial relationship between the government and employers. The justice system has attempted to provide some relief by reducing the burden of the injured, but this is not the answer. Rising workmen's compensation costs have created some incentive for employers to become self-insured and to reduce costs by focusing on hazard prevention. Management and operating personnel need more information on the hazards that repeatedly cause injury so they can optimize production through hazard prevention. Most people, when made aware of hazardous circumstances, will be very responsive. Doing nothing to prevent inherent crane hazards is morally unacceptable. System safety fosters continuous design improvement to overcome human performance variables. Weak standards must go. We must upgrade minimum standards to assure that our workers are provided safe equipment in a safe workplace. The burden placed on the community when it must assure for the welfare and care of the injured and their dependents has created a whole new set of safety economics. The prevention of hazards is clearly cost-effective and creates more wealth for all. Ignoring injury costs has led to skyrocketing insurance premiums that dwarf the minute cost of life-saving design improvement.

The National Commission for the Certification of Crane Operators is becoming an effective means of upgrading crane operator performance.

National Commission for the Certification of Crane Operators
2750 Prosperity Avenue, Suite 505
Fairfax, Virginia 22031-4312
Telephone: (703) 560-2391
Fax: (703) 560-2392
E-Mail: info@nccco.org
Executive Director: Graham Brent

Operator certification gives the crane operator greater authority at the worksite to ensure for a safe set-up, which allows him or her to decline the use of the crane in unsafe circumstances or if the crane is in unsafe condition. Like the captain of a ship, a certified crane operator has the last clear opportunity to refuse work in dangerous locations (such as next to powerlines), use the crane in a dangerous manner (such as outriggers retracted) or use a crane which is unfit (such as a crane in disrepair).

Over the last few years, this commission has made an outstanding contribution to the safe operation of cranes. As technology continues to provide the blessing of operator aids, the load moment indicator (LMI) is significantly reducing the occurrence of crane upset from overloads. However, too many crane operators are intimidated by this device and attempt to bypass it. Certification that an operator is skilled in the use of an LMI would be a further enhancement of this program. When a crane equipped with an LMI is brought onto a job site, both the construction manager and the safety director need to know whether the operator knows how to use this device.

3

Crane Types

1. Mobile Hydraulic Cranes and Wheel-Mounted Telescoping Boom[1]

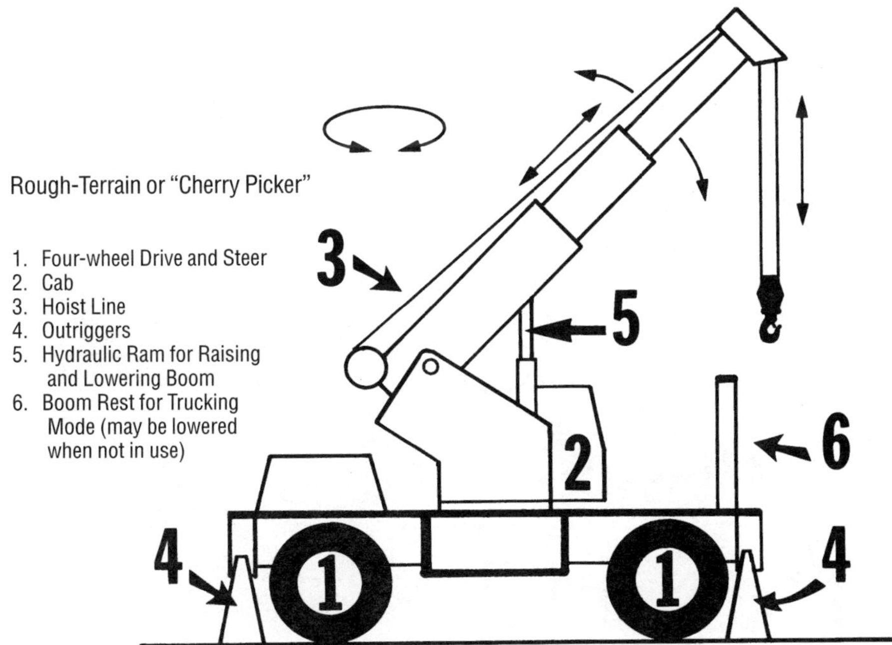

| *Figure 3-1* Telescoping Boom Mobile Hydraulic Crane |

---

[1] Pertinent ANSI standards provide additional detailed illustrations of the various crane configurations.

2. Truck-Mounted Cranes

Figure 3-2 **Truck-Mounted Hydraulic Boom Crane**

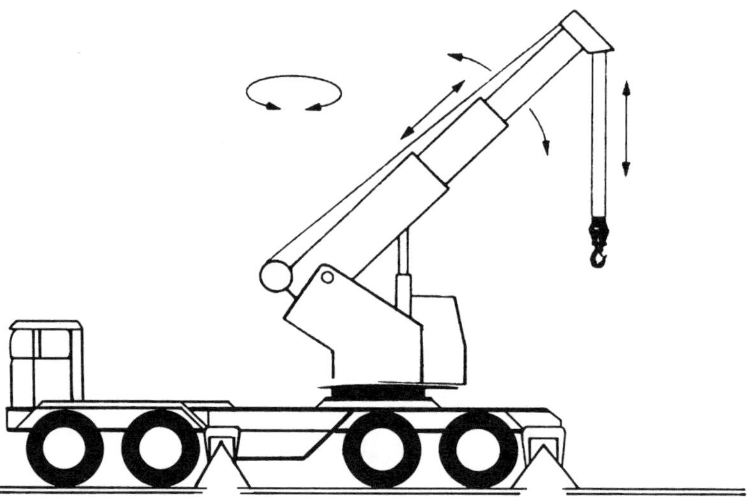

Figure 3-3 **Latticework Fixed-Boom Crane**

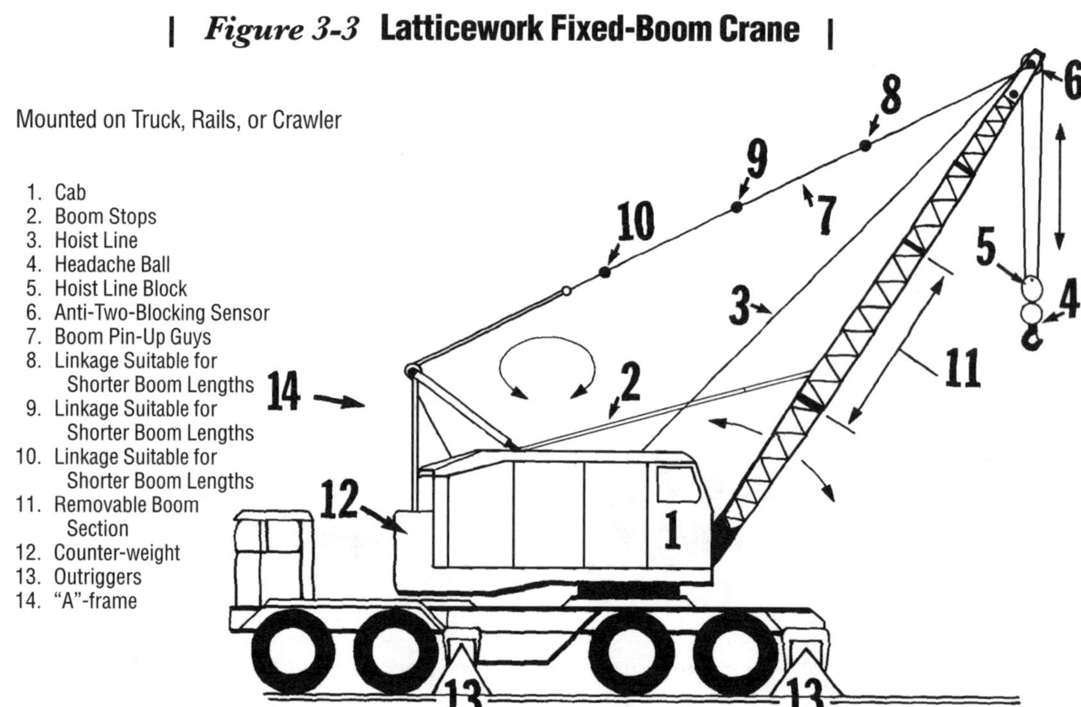

3. Flatbed-Truck Mounted Cranes

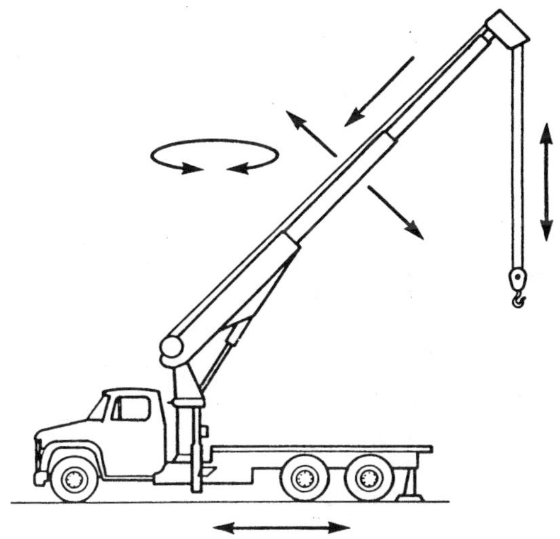

| *Figure 3-4* **Pedestal Hydraulic Boom** |

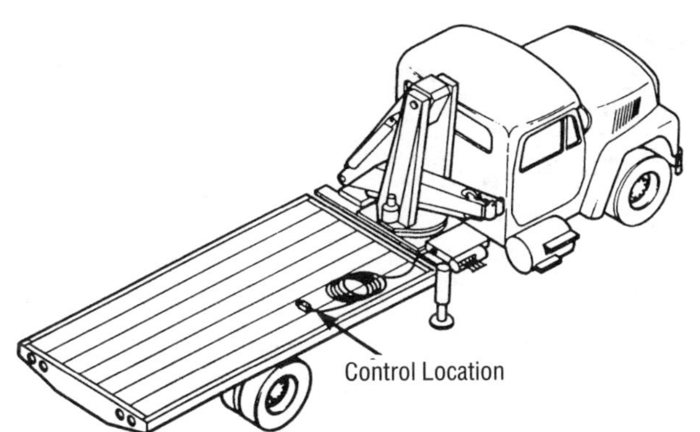

| *Figure 3-5* **Articulated Boom** |

ANSI B30.22–1981
Authored by ASME

| *Figure 3-6* **Trolley Boom** |

4. Overhead-Track Mounted Cranes

Figure 3-7A Overhead-Track Mounted Cranes

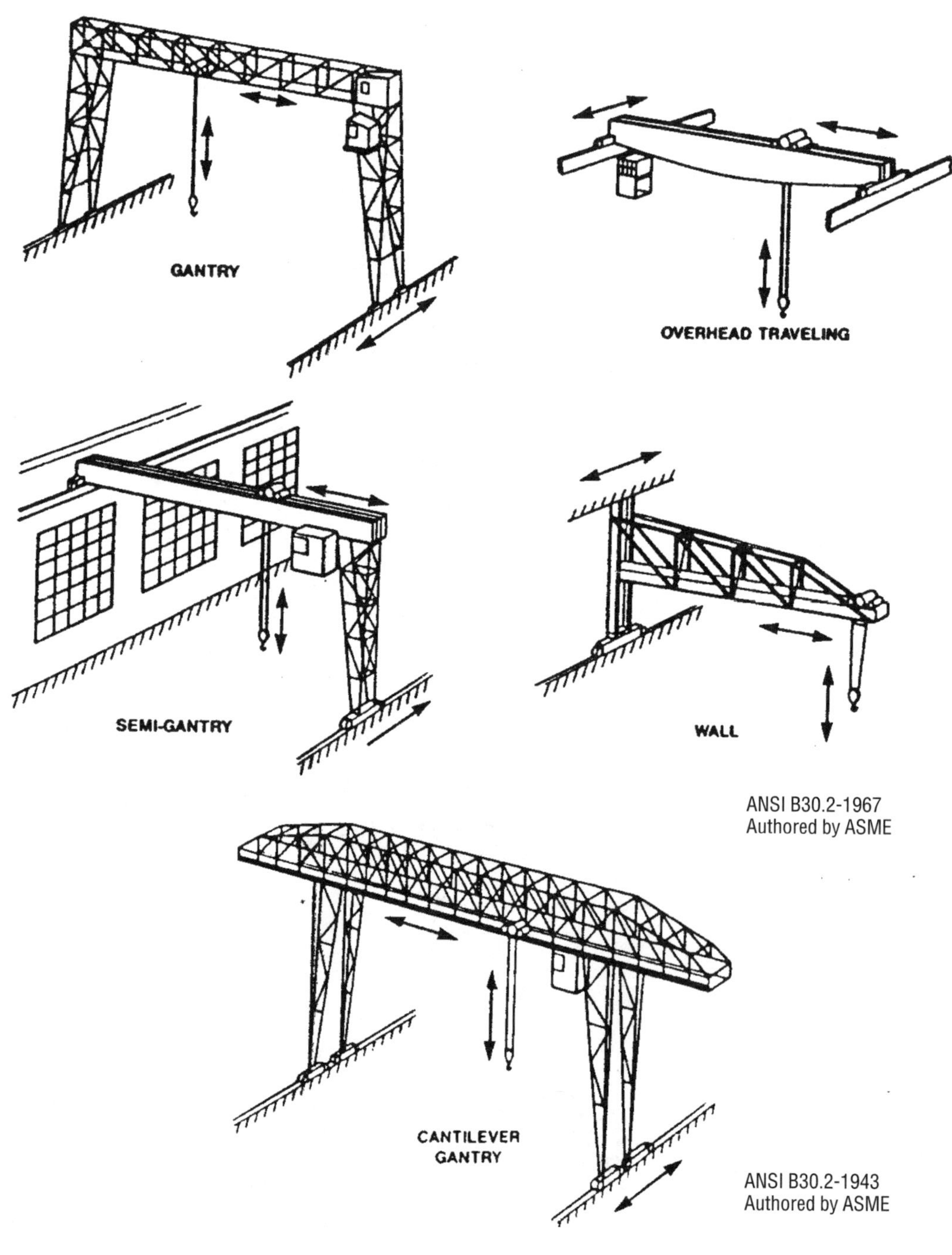

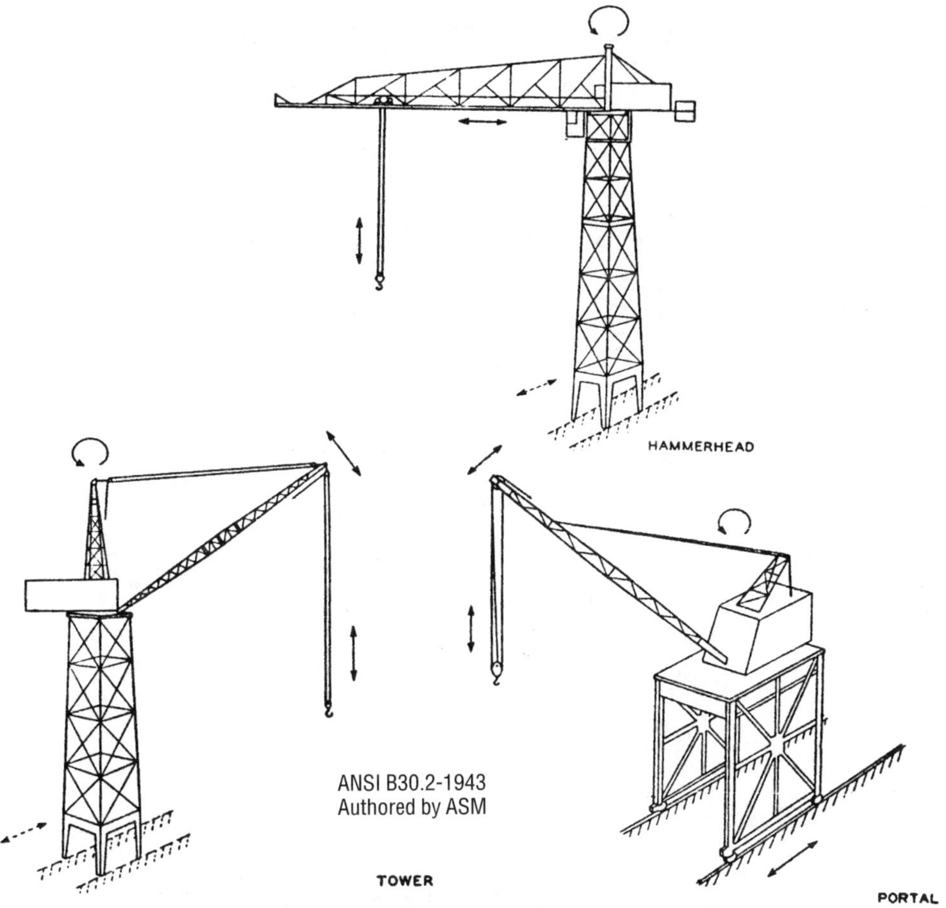

Figure 3-7B **Track-Mounted Fixed-Boom Cranes**

5. Monorails and Underhung Cranes (Hoist)

Figure 3-8 **Examples of Monorails and Underhung Hoists**

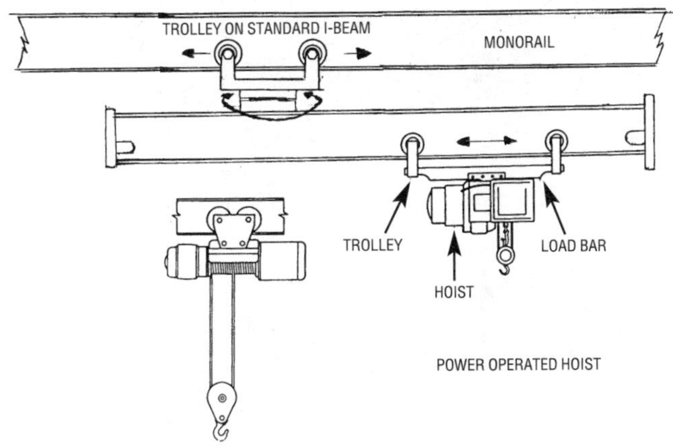

7. Straddle Cranes (Lift)

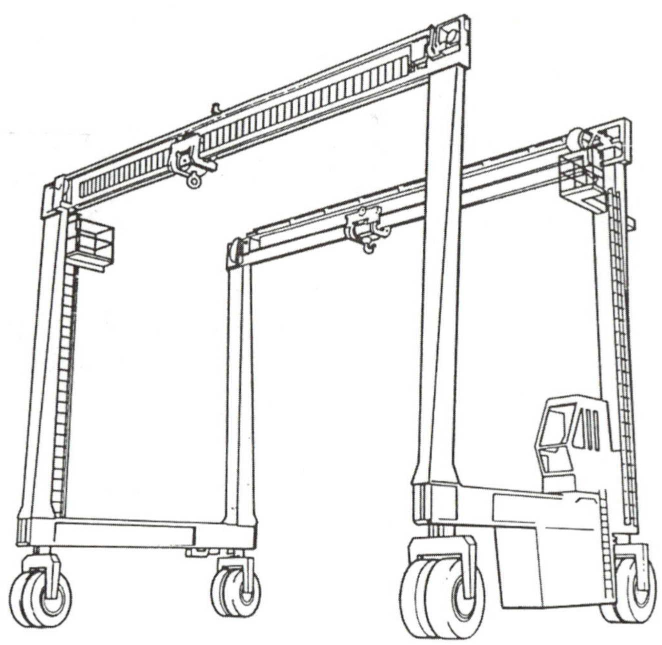

Figure 3-9 Wheel-Mounted Straddle Lift

8. Fixed Cranes

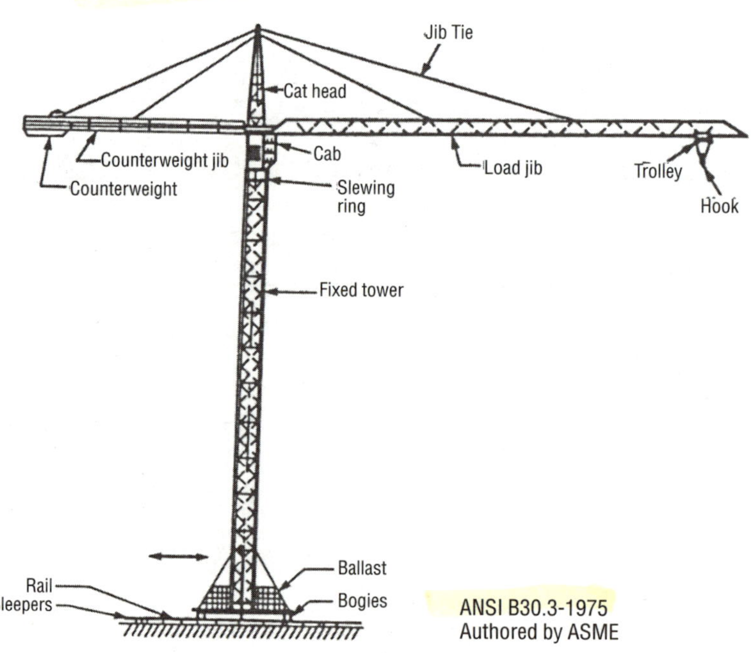

Figure 3-10 Hammerhead Tower Cranes

| *Figure 3-11* **Derrick Crane** |

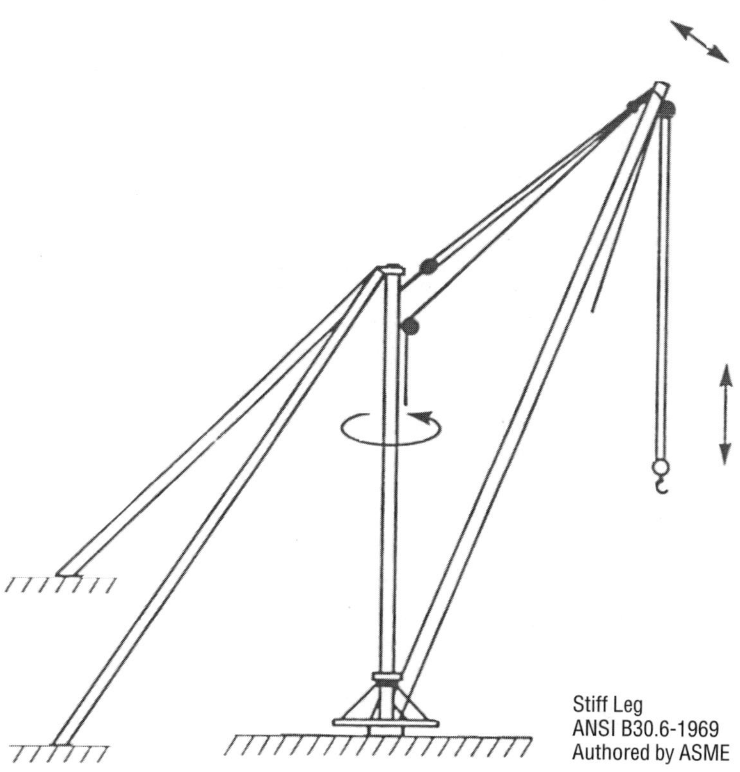

| *Figure 3-12* **Digger Derrick** |

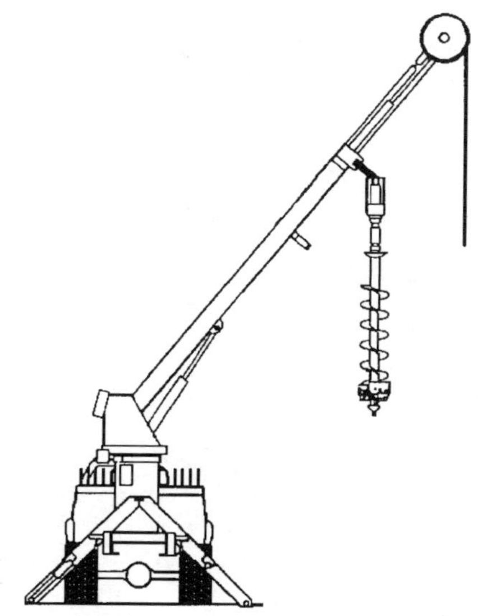

Figure 3-13 Aerial Lift and Examples of A92.5 Aerial Platforms
(Figures courtesy of ANSI/SIA.)

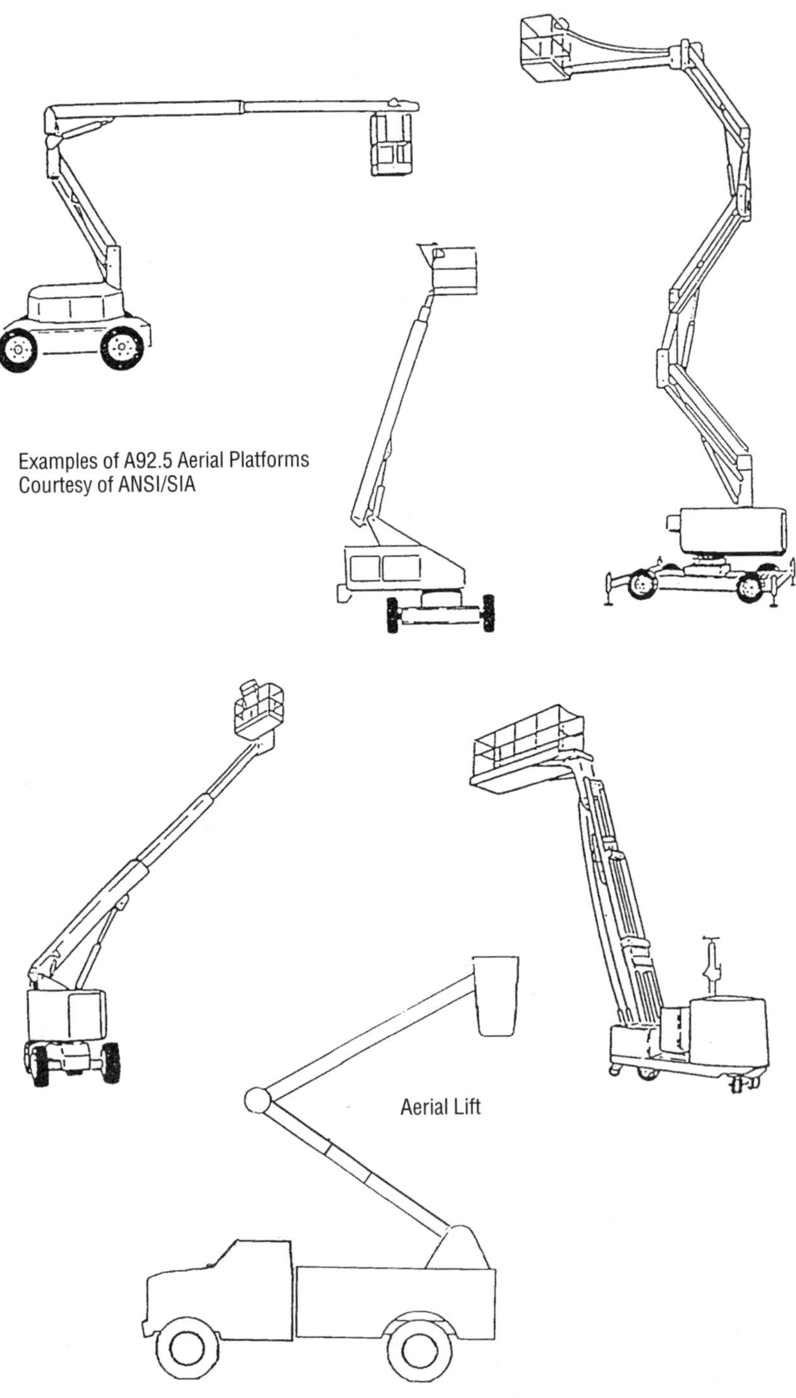

4

Powerline Contact

Powerline Contact by Hoist or Boom

Definition

Powerline contact occurs when any metal part of a crane touches a bare, uninsulated, high-voltage powerline.

Description

Most powerline contacts occur when the crane's hoist line or boom touches a powerline during the moving of materials adjacent to or under energized powerlines. Contact also frequently occurs during pick-and-carry operations when loads are being transported under energized powerlines. Sometimes the victim of electrocution has been touching the crane or getting on or off the crane without knowing that the hoist line or boom has inadvertently come into contact with an energized powerline.

In some circumstances, when sufficient ground fault is created, the electric distribution circuit is automatically de-energized by a remote re-closure system to avoid the blowing of intervening fuses, with the circuit being automatically re-energized several seconds later. Victims are often re-shocked and rescuers endangered because they believe the powerline has been permanently de-energized and do not know that the re-closure system was re-energized after only a few seconds. Where cranes are working in the vicinity of powerlines, it is good practice to have the utility company lock out the automatic re-closures so the line remains de-energized and is not re-energized a second or two later.

Risks Presented by Hazard

Cranes and powerlines are not compatible and should not occupy the same work space. Powerline contact presents the highest risk in crane operations. It is probably the most devastating, continually reoccurring type of personal injury and property damage. A single contact can result in multiple deaths and/or crippling injuries. From my review of OSHA and MSHA injury data and injury data available from litigation experience, my best estimate shows that each year approximately 150 to 160 people are killed or crippled by powerline contact and about three times that number are seriously injured. On an average, eight out of ten of those injured were guiding the load at the time of contact.

Available Hazard Prevention Measures

The occurrence of cranes or other boomed equipment contacting powerlines is generally the result of a complex involvement of multiple parties who could have all interceded on an individual basis and prevented the event. The knowledge of all the facts that provided the opportunity for a crane or other boomed equipment to contact the powerline needs to be researched in order to identify the responsible parties. The key to the prevention of powerline contacts is to eliminate and minimize the potential of a powerline contact before the crane and crew arrives at the work site. In March 2004, the Hazard Information Foundation, Inc. (HIFI) published a 230-page research report entitled "Safety Interventions to Control Hazards Related to Powerline Contacts by Mobile Cranes and Other Boomed Equipment" that was co-funded by the Center to Protect Workers' Rights (CPWR). To obtain a copy of this report, contact HIFI by email at besafe@hazardinfo.com or by telephone at (520) 458-6700. This report highlights a time-line history of over 50 years of significant studies, reports, and references concerning powerline contact and contains fifty in-depth summaries of powerline contact occurrences to illustrate the various types of crane, aerial lifts, and other boomed equipment involved in powerline contacts. These portions of the research study provide a basis for 34 recommendations. Overall, the recommendations illustrate how the management of various parties can initiate simple actions to prevent the conditions and circumstances that lead to powerline contact. For safety training to crane operators and users of cranes it is essential that management of electric utilities, facility developers and landowners, construction managers, general and subcontractors, crane and lift manufacturers, equipment dealers and rental agencies all play their important safety role to eliminate or minimize the hazard of powerline contact by the design and/or use of safety appliances.

The key to avoiding powerline contact is pre-job safety planning. (See pre-job planning references, bibliographic listings 108-119.) Planning is one of the greatest deterrents to avoiding injury or property damage available in the workplace. Because of the large number of parties and employers involved in controlling the construction worksite (the landowner, construction management, prime contractor, subcontractors, crane rental firms, electric utilities, etc.), planning is a necessity to establish who is in charge. A single individual should take on overall safety supervision and coordination of the project, and this individual must initiate positive steps to assure that pre-job safety planning has been completed before any cranes arrive at the worksite. The objective of this pre-construction planning is to organize the work so that the crane operations will not intrude in any areas where powerlines are located or will be erected. Site plans must show the location of existing or planned powerlines, so existing powerlines can be rerouted before any cranes are brought onto the worksite or so the area can be fenced and declared off-limits to cranes. Location of powerlines on site plans will also assist in locating routes for pick-and-carry and other crane travel to avoid having cranes pass under powerlines. Pre-construction planning must also consider the clearances of existing powerlines, as these clearances might be acceptable for normal roadway traffic, but not for cranes. In Table 232-1 of the *National Electrical Safety Code* (Figures 4-1A, 4-1B and 4-1C), and Request #IR 159 dated April 11, 1974, and its Interpretation in *National Electrical Safety Code Interpretations, 1961-1977 Inclusive* (Figure 4-2), it is clearly set forth that the clearances listed are not applicable to construction, maritime activities, mining, etc. where cranes may be used.

Contractors are required to inform the local electric utility that cranes are to be used in specific areas of the work site. If crane operations can penetrate the danger zone surrounding powerlines, the electric utility must be notified to de-energize, relocate, bury, or insulate the powerlines before the crane is operated in that location. (See OSHA 1910.180(j), "Operating Near Electric Powerlines" and OSHA 1926.550(a)(15).) If the utility has not done any of the above, it should be asked to lock out the automatic re-closure system. (See OSHA 1926.955(e)(5).) The electric utility has the ultimate responsibility for placement of powerlines outside of harm's way and plays an important role in pre-job planning. It should be a strong advocate for temporarily relocating powerlines outside the reach of cranes during the entire construction process from site preparation to completion. To be outside the reach of cranes, powerlines must be located ten feet beyond the 360 degree sluing radius of a boom in the horizontal position. This requirement ensures that no part of the boom or hoist line can be brought into contact with a powerline. The electric utility is entitled to be reimbursed for undertaking these safety measures, and this cost should be included in the project budget.

Owners of the property which has powerlines on their property should not direct or allow work with boomed equipment to be conducted when the powerline can be inadvertently struck. It must be considered that such practices are inherently dangerous and can predictably lead to serious injury or death.

For years, human factors specialists have stated that it is beyond the range of normal human performance to:

1. accurately, visually judge clearances between a crane boom or hoist line and powerlines

2. observe more than one visual target at a time

3. overcome the camouflaging characteristics that trees, buildings, and other objects have upon powerlines

Sole reliance upon the performance of crane operators, riggers, and signalers, without mapping the danger zone to specify crane/powerline clearance has resulted in many deaths. Without identifying the danger zone on the ground with marker tape or other visible clues, visual judgment of overhead clearance can be in error if the spotters are not properly positioned or are momentarily distracted. Safety planning to avoid the use of cranes adjacent to powerlines is the most reliable method of avoiding crane/powerline contacts. Because important considerations can be overlooked in pre-job planning, it is also extremely important to provide training for crane operators, riggers, and supervisors on how to map the danger zone on the ground.

If powerlines have not been relocated or buried and crane operations are to be carried on adjacent to them, the crane operator, those guiding the load, and those closely involved in the operation need some visual guidance on the ground to make them aware of the danger zone. The area within a radius of ten feet in any direction from powerlines is an unsafe work area and must be clearly marked off on the ground by marker tape, fences, barriers, etc. Everyone at the worksite must be sure that the crane is positioned so the boom or the hoist line cannot intrude into this danger zone during sluing or other operations. Job-site foremen, operating personnel, riggers, crane operators, and signalers must be trained to map the danger zone. The easiest way to identify the parameters of the danger zone is to step off fifteen feet on each side of each power pole in the work area, draw a continuous line from pole to pole, and place some form of visual barrier on the line so that any encroachment of the hoist line, boom tip, or load into this danger zone will be visible to all personnel involved in the crane operations. Figures 4-3A, 4-3B, and 4-3C show how this mapping is done so that no part of the crane intrudes into the danger zone.

Unintentional powerline contact often occurs during pick-and-carry operations by mobile cranes traveling a previously traversed route. OSHA *Code of Federal Regulations,* Subpart N, 1926.550(a)(15)(v) lists cage-type boom guards, insulating links, and proximity warning devices solely as redundant back-up measures and cautions that they are not substitutes for maintaining a safe clearance. These devices, if used in accordance with manufacturers' recommendations, have successfully functioned as additional safeguards for pick-and-carry operations, as they provide an electrical barrier or warning of danger. Their use also creates an ongoing awareness of the danger and fosters an increased effort to avoid exposure to this life-threatening hazard. Insulated links (see Figure 4-4) are back-up devices that prevent the flow of current down the hoist line and protects the individual guiding the load. The insulated link serves as a redundant safeguard in the event that the powerline was not observed. Dr. George Karady of Arizona State University has tested insulated links and other insulating materials in their protective capacity, even when contaminated by foreseeable materials, and his tests have all indicated even the contaminated links are able to provide lifesaving protection. Tests of insulated links under field conditions by Martin N. Kaplan, P.E., show that these safeguards have a positive ability to save lives, even when they have been thoroughly covered with mud, oil, and other contaminants common to the work area. Kaplan's tests clearly demonstrated that any leakage of amperage was well below the paralysis threshold and so indicated that life-threatening exposure to riggers touching the load can be eliminated. Without an insulated link, riggers are in the ground fault circuit if powerline contact is made. Further, Kaplan's tests assumed a more conservative threshold for "dangerous" current than OSHA's general requirements that ground fault interrupters operate at five milliamperes. Note: the Army and Navy have both used insulated links and have developed confidence in them.

Over the years there has been considerable debate concerning the effectiveness of proximity alarms. Energized powerlines create two distinct measurable signals — (1) electrostatic, which is based upon the voltage of the powerline, and (2) magnetic, which is based upon the current flow. Voltage produces a constant signal. Current flow varies from time to time, making the signal unreliable. For these reasons, a purchaser of a proximity alarm should select one that is certified by the manufacturer to rely upon static voltage to sound an alarm when a boom is approaching a powerline. Today a number of firms have used the electrostatic proximity alarm on a number of cranes for a number of years, and their experience has been that the device provides a high level of awareness and ensures avoidance of powerlines.

My research shows that inadvertent powerline contact by truck-mounted trolleys or articulated crane booms that utilize electrical remote-control systems to load or unload bricks, cement blocks, trusses, or other building supplies has also caused much injury and death. If the boom or trolley contacts a powerline, the person using the control box at the end of the electrical control cable is usually instantly electrocuted. Control boxes of this type should never be used under powerlines. A safer design choice would be non-conductive, pneumatic-powered control systems or remote radio control systems. (See Chapter 10 for additional information.)

Another device that has aided crane operators in avoiding powerline contacts is a range-limiting device. This device is required in the Canadian province of Quebec in its *Safety Code for the Construction Industry*, S-2.a, r. 6, paragraph 5.2.2(c). The distributor for this device in the United States is Wylie Systems of Tulsa, Oklahoma. It markets this device under the trade name Boom Buoy. This device allows the crane operator to program safe clearances when working adjacent to powerlines by giving the operator the opportunity to program a safe working area of invisible boundaries on all sides and above to prevent raising or sluing of the boom into a powerline. It also gives visual and audible warning when the crane's boom encroaches upon these invisible boundaries.

Pedestal cranes mounted on a flatbed truck should not have controls located so a person standing on the ground can reach them, as this leaves the operator vulnerable to the initial fault current path if a boom happens to strike a powerline.

OSHA Requirements

1. OSHA 1926.550(a)(15).
2. OSHA 1926.955(e)(5) states: "The automatic re-closing feature of circuit interrupting devices shall be made inoperative where practical before working on any energized line or equipment."
3. Marine Terminal Standard 1917.45(i)(5).

ANSI Requirements

1. *National Electrical Safety Code*, ANSI C2, Table 232 with Footnotes (see Illustrations Nos. 13A, 13B, and 13C), and Section 20, Paragraphs 200-214.
2. *Mobile and Locomotive Cranes*, ANSI/ASME B30.5-1994, B30.5-2000 section 5-3.4.5.1 through .4 and figures 17(a), (b), (c) and (d), and Addendum 1995 Section 5-3.4.5.1 through .4. See figures 17a and 17b.
3. *Mobile Hydraulic Cranes*, ANSI B30.15, Section 15-3.4.2.
4. *Articulating Boom Cranes*, ASME/ANSI B30.22, Section 22-3.3.

Other References

1. *Safety and Health Requirements Manual*, EM 385-1-1 (U.S. Army Corps of Engineers, October, 1992), Section 11.E.01.b. states: "All electric power or distribution lines shall be placed underground in areas where there is extensive use of equipment having the capability to encroach on the clear distances specified in 11.E.04.
2. *Guidance Note General Series 6*, "Avoidance of Danger from Overhead Electric Lines," HM Factory Inspectorate of the Health and Safety Executive and the Electricity Boards of England, Wales and Scotland (April, 1977).
3. United Kingdom Regulation 44 of

the *Construction* (General Provisions) *Regulations* of the Factories Act of 1961 stipulates that powerlines left in the construction area be barricaded to provide a wooden goal post framework to prevent the crane from striking a powerline.

4. *Crane Handbook* by D. E. Dickie of the Canadian Construction Safety Association of Ontario (1975), pp. 133-141, and Figures 4.66-477.

5. *Mobile Crane Manual* by D. E. Dickie of the Canadian Construction Safety Association of Ontario (1982), pp. 240-244, and Figure 10.15.

Suggested Design Criteria

Electric utilities should be asked to remove powerlines from the area where cranes will be used at the contractor's expense. Utilities should provide a brochure for contractors and other crane users on how to map the danger zone, with specific instructions to notify the utility whenever it is found that crane operations will encroach into the danger zone. The brochure should provide a specific telephone number and the name of the person to contact for assistance. The electric utility should recognize that contractors or construction management may be uninformed as to the limits of human performance in estimating visually safe clearances; therefore, powerlines must be removed from crane operations as a reasonable and prudent workplace safety requirement.

Operator's manuals should incorporate mapping instructions for avoidance of powerlines (Figures 4-3A, 4-3B, and 4-3C). Also, these manuals should indicate, as does OSHA, that insulated links and proximity alarms are available as redundant back-up measures, but not as substitutes for safe clearance.

Representative Litigation

As identified by HIFI:

- Mobile hydraulic telescoping boom cranes are the most frequent equipment to contact powerlines.
- Flat-bed mounted pedestal hydraulic telescoping boom cranes are the second most frequent type of crane contacting powerlines.
- Aerial lifts, uninsulated and insulated, are the third most frequent type of boomed equipment to contact powerlines.
- Latticework boom cranes are the 4th most frequent type of equipment to contact powerlines.
- Pumpcrete booms, straddle cranes, knuckle booms, shingle conveyors, drill rigs, and dump trucks all share a tragic record of powerline contact.

The HIFI study included 50 detailed descriptions of powerline contact cases that are representative of over 2,000 known powerline contact occurrences. Due to varying statistical thoroughness, it is difficult to assign a frequency of fatalities to powerline contacts. The best estimate is one fatality for every five powerline contacts that result in serious injury and the basis for personal injury liability. The casualty rate of the study's 50 cases was much higher, as it discussed occurrences that were very severe and resulted in 25 deaths and 42 crippling injuries.

When examining the issue of powerline contact, facts provided by litigation are the best source of information on what can be done by management before the boomed equipment and workers arrive at the work site. Attempting to control the hazard by delegating the responsibility for prevention to operating personnel is only a stop-gap measure for management's failure to eliminate this serious hazard in the first place.

This litigation data could be used to show management the benefits of expanding the responsibilities of its safety department to develop programs to target this critical hazard. A real need exists for utility companies to develop training programs for contractors, crane rental firms, and crane users on how to map the danger zone as discussed in Chapters 4, 10 and 19. To avoid repetitive litigation involving boomed equipment such as cranes, aerial baskets, high-reach lifts, and pumpcrete machines, utility companies should be involved with and promote pre-construction planning to avoid use of boomed equipment adjacent to energized powerlines. The insistence on such planning would provide utilities with a valid, written record proving it had gone the extra mile by informing and training those managing and directly responsible for the use of cranes in dangerous locations next to powerlines.

Another factor that becomes apparent from examination of these summaries is the absence of information in electric utility easements that warns the landowner: 1) that overhead powerlines limit the use of the land covered by the easement, 2) that this space should never be used for storage of materials, and 3) that cranes and other boomed equipment should never be used adjacent to powerlines. A written easement should also contain specific instructions on how to identify the danger zone and to notify the utility to de-energize the powerlines before any project is undertaken that might encroach upon this area.

A more detailed analysis of the types of cranes involved in powerline contacts is:

1. Truck Carrier, Latticework Boom: 26%
2. Truck Carrier, Hydraulic Boom: 24%
3. Mobile Hydraulic Boom, Rough-Terrain: 19%
4. Flatbed, Hydraulic Boom: 16%
5. Flatbed, Trolley Boom, Remote Control: 11%
6. Crawler Carrier, Latticework Boom: 4%
7. Large latticework cranes are known to slew or back the pennant support system into powerlines

The personnel were injured or killed when:

1. Guiding the load: 71%
2. Getting on or off the crane and/or touching the crane: 21%
3. Other activity: 8%

Eighty percent of the time the voltage of the powerlines contacted was less than 16,000 volts. Twenty percent of the time the voltages were more than 16,000 volts. It can be easily concluded that insulated links rated for 50,000 volts (a safety factor of 3) and tested for 100,000 volts (a safety factor of 6) could reduce the injury rate from electrocution by at least sixty-four percent. Insulated links should be provided as redundant, back-up safety devices. In pick-and-carry and other operations, the use of a proximity alarm is a helpful reminder that a powerline is in the path of travel. In many powerline contact cases, the use of insulated links and proximity alarms is advocated as a hazard prevention measure. There is no record of injuries resulting from the use of these devices. Substantial testimony has been given by U.S. Navy personnel and others that insulated

links and proximity alarms are effective back-up devices, but, more importantly, they create a strong awareness in personnel of the need to avoid powerlines in crane operations. Some defense attorneys attempt to excuse their client's liability for failing to use insulated links and proximity alarms by retaining paralogists to disparage the reliability of these devices.

Representative Cases

Plaintiff Iniguez; occurred April 23, 1968; Superior Court, Maricopa County, Arizona #C-228130.

> Injured lost arm when removing hook when placing four-foot concrete drainage culvert in ditch. Boom popped up into 7,200-volt powerline when the load was released. Alleged utility was aware of the danger and failed to de-energize the powerline. Verdict for the injured.

Plaintiff Burke; occurred July 20, 1970; Appellate Court of Illinois, First Judicial District, Third Division, #59549.

> One worker killed, one permanently disabled when they were guiding pipe as sluing boom hoist line struck 34,500-volt powerline. Alleged the electric utility did not de-energize powerline and the crane manufacturer did not propose use of insulated link and proximity alarm. Verdict for the injured.

Plaintiff Gordon; Filed 1973; Fourth District Court, Alaska, #73-986.

> Oiler guiding load sustained serious injury when backhoe boom touched 7,200-volt single-phase powerline with neutral below, while placing drainage system in housing development. Alleged utility failed to de-energize line while work was in progress. Work area intruded powerline danger zone. Pre-trial settlement.

Plaintiff Gordon; occurred January 30, 1975; Henry County Circuit Court, Indiana, #77-C-470.

> Foreman lost feet and hands while guiding materials being unloaded from truck. Hoist line contacted

| *Figure 4-1A* **National Electrical Safety Code Table 232-1** |

Table 232-1. Minimum Vertical Clearance of Wires, Conductors, and Cables Above Ground, Rails, or Water FT

(Voltages are phase to ground for effectively grounded circuits and those other circuits where all ground faults are cleared by promptly de-energizing the faulted section, both initially and following subsequent breaker operations. See the definition section for voltages of other systems.)

Nature of surface underneath wires, conductors, or cables	Communication conductors and cables, guys, messengers, surge protection wires, neutral conductors meeting Rule 230E1, supply cables meeting Rule 230C1 and supply cables of 0 to 750 V meeting Rules 230C2 or 230C3 (11) (ft)	Open supply line conductors of 0 to 750 V. and supply cables over 750 V meeting Rule 230C2 or 230C3 (ft)	Open supply line conductors 750 V to 22 kV (ft)	22 to 50 kV (ft)	Trolley and electrified railroad contact conductors and associated span or messenger wires (1) — 0 to 750 V to ground (ft)	750 V to 50 kV to ground (ft)
Where wires, conductors, or cables cross over or overhang						
1. Track rails of railroads (except electrified railroads using over-head trolley conductors) (2)(16)(20)	(3)(15) 27	(3)27	(3)28	29	(4)22	(4)22
2. Roads, streets, alleys; nonresidential driveways, parking lots, and other areas subject to truck traffic (21)(22)	(6)(13)(23)18	18	20	21	(5)18	(5)20
3. Residential driveways; commercial areas not subject to truck traffic (21)(22)	(24)12	(8a) 15	20	21	(5)18	(5)20
4. Other land traversed by vehicles such as cultivated, grazing, forest, orchard, etc	18	18	20	21	—	—
5. Spaces or ways accessible to pedestrians only (9)	(8)(7) 15	(8a)(14)15	15	16	16	18
6. Water areas not suitable for sailboating or where sailboating is prohibited (19)	15	15	17	17	—	—
7. Water areas suitable for sailboating including lakes, ponds, reservoirs, tidal waters, rivers, streams, and canals with an unobstructed surface area of: (17)(18)(19)						
(a) Less than 20 acres	18	18	20	21	—	—
(b) 20 to 200 acres	26	26	28	29	—	—
(c) 200 to 2000 acres	32	32	34	35	—	—
(d) Over 2000 acres	38	38	40	41	—	—
8. Public or private land and water areas posted for rigging or launching sailboats	Clearance above ground shall be 5 ft greater than in 7 above, for the type of water areas served by the launching site					
Where wires, conductors, or cables run along and within the limits of highways or other road rights-of-way but do not overhang the roadway						
9. Roads, streets, or alleys	(13)(23)(25) 18	18	20	21	(5)18	(5)20
10. Roads in rural districts where it is unlikely that vehicles will be crossing under the line	(10)(12) 14	(10) 15	18	19	(5)18	(5)20

| *Figure 4-1B* **Table 232-1 Footnotes** |

Footnotes for Table 232-1.   **FT**

① Where subways, tunnels, or bridges require it, less clearances above ground or rails than required by Table 232-1 may be used locally. The trolley and electrified railroad contact conductor should be graded very gradually from the regular construction down to the reduced elevation.

② For wire, conductors, or cables crossing over mine, logging, and similar railways which handle only cars lower than standard freight cars, the clearance may be reduced by an amount equal to the difference in height between the highest loaded car handled and 20 ft, but the clearances shall not be reduced below that required for street crossings.

③ These clearances may be reduced to 25 ft where paralleled by trolley-contact conductor on the same street or highway.

④ In communities where 21 ft has been established, this clearance may be continued if carefully maintained. The elevation of the contact conductor should be the same in the crossing and next adjacent spans. (See Rule 289D2 for conditions which must be met where uniform height above rail is impractical.)

⑤ In communities where 16 ft has been established for trolley and electrified railroad contact conductors 0 to 750 V to ground, or 18 ft for trolley and electrified railroad contact conductors exceeding 750 V, or where local conditions make it impractical to obtain the clearance given in the table, these reduced clearances may be used if carefully maintained.

⑥ If a communication service drop or a guy which is effectively grounded or is insulated against the highest voltage to which it is exposed, up to 8.7 kV, crosses residential streets and roads, the clearance may be reduced to 16 ft at the side of the traveled way provided the clearance at the center of the traveled way is at least 18 ft. This reduction in clearance does not apply to arterial streets and highways which are primarily for through traffic, usually on a continuous route.

⑦ This clearance may be reduced to the following values: *feet*

  (a) For insulated communication conductors and communication cables   8
  (b) For conductors of other communication circuits   10
  (c) For guys   8
  (d) For supply cables meeting Rule 230C1   10

Footnotes for Table 232-1
Continued on pages 145 and 146.

⑧ This clearance may be reduced to the following values:

  (a) 12 ft for supply conductors limited to 300 V to ground
  (b) 10 ft for drip loops of service drop conductors limited to 150 V to ground and meeting Rules 230C2 or 230C3 and the portion of the associated service drop span located within 15 ft of the service entrance to buildings.

⑨ Spaces and ways accessible to pedestrians only are areas where vehicular traffic is not normally encountered or not reasonably anticipated.

⑩ Where a supply or communication line along a road is located relative to fences, ditches, embankments, etc, so that the ground under the line would not be expected to be traveled except by pedestrians, this clearance may be reduced to the following values:

                                    feet

  (a) Insulated communication conductor and communication cables   8
  (b) Conductors of other communication circuits   10
  (c) Supply cables of any voltage meeting Rule 230C1 and supply cables limited to 150 V to ground meeting Rules 230C2 or 230C3   10

  (d) Supply conductors limited to 300 V to ground   12
  (e) Guys   8

⑪ No clearance from ground is required for anchor guys not crossing track rails. streets, driveways, roads, or pathways.

⑫ This clearance may be reduced to 13 ft for communication conductors.

⑬ Where this construction crosses over or runs along alleys, driveways, or parking lots, this clearance may be reduced to 15 ft for spans limited to 150 ft.

⑭ Where supply circuits of 600 V or less, with transmitted power of 5000 W or less, are run along fenced (or otherwise guarded) private rights-of-way in accordance with the provisions specified in Rule 220B2, this clearance may be reduced to 10 ft.

⑮ The value may be reduced to 25 ft for guys, for cables carried on messengers, and for supply cables meeting Rule 230C1. This value may be reduced to 25 ft for conductors effectively grounded throughout their length and associated with supply circuits of 0 to 22 kV, only if such conductors are stranded, are of corrosion-resistant material, and conform to the strength and tension requirements for messengers given in Rule 2611.

| *Figure 4-1C* **Table 232-1 Footnotes (cont.)** |

(16) Adjacent to tunnels and overhead bridges which restrict the height of loaded rail cars to less than 20 ft, these clearances may be reduced by the difference between the highest loaded rail car handled and 20 ft, if mutually agreed to by the parties at interest.

(17) For controlled impoundments, the surface area and corresponding clearances shall be based upon the design high water level. For other waters, the surface area shall be that enclosed by its annual high water mark, and clearances shall be based on the normal flood level. The clearance over rivers, streams, and canals shall be based upon the largest surface area of any 1 mi long segment which includes the crossing. The clearance over a canal, river, or stream normally used to provide access for sailboats to a larger body of water shall be the same as that required for the larger body of water.

(18) Where an overwater obstruction restricts vessel height to less than the following:

For a surface area in acres of	A reference vessel height in feet of
less than 20	16
20 to 200	24
200 to 2000	30
over 2000	36

the required clearance may be reduced by the difference between the reference vessel height given above and the overwater obstruction height, except that the reduced clearance shall not be less than that required for the surface area on the line crossing side of the obstruction.

(19) Where the US Army Corps of Engineers, or the State, or a surrogate thereof has issued a crossing permit, clearances of that permit shall govern.

(20) See Rule 234H for the required horizontal and diagonal clearances to rail cars.

(21) These clearances do not allow for the future road resurfacing.

(22) For the purpose of this rule, trucks are defined as any vehicle exceeding 8 ft in height. Areas not subject to truck traffic are areas where truck traffic is not normally encountered or not reasonably anticipated.

(23) For communications cables supported on a messenger, and with span lengths not exceeding 150 ft, the clearance may be reduced to 17 ft above or along local streets or roads. This reduction does not apply for arterial streets or highways which are primarily for through traffic, usually on a continuous route.

(24) This clearance may be reduced to 10 ft for communication conductors and cables, guys, messengers and supply cables meeting Rule 230C1.

(25) Communication cables supported on a steel messenger may have a 60°F clearance of 15 ft where span lengths do not exceed 150 ft and poles are back of curbs or other deterrents to vehicular traffic.

| *Figure 4-2* **Request #1R 59 and its Interpretation in NESC** |

232A Table 1   55   232A Table 1

232A Table 1
See IR for Rule 230D, IR 126(b)

Clearances applicable to building construction site

REQUEST (Apr 11, 74)   IR 159

Can you advise whether there has been an interpretation of . . . Rule 232A to determine under what category a building construction site would fall? The building would be multistory, height exceeding the maximum clearances shown in Table 1 of Rule 232A. Scaffolding, ladders, fork-lift trucks, and other vehicles and equipment would be expected to be used in the area.

INTERPRETATION (Oct 7, 74)

Clearances involving building construction sites are not covered by this Code. This was the subject of a previous interpretation dated March 12, 1963 (rule 234C4, IR 98). It was pointed out that there are too many variables which may affect clearance requirements for buildings under construction. It was felt that any set of clearances which covered the worst situation would inherently penalize other situations where different construction equipment and methods were used. Clearances from completed buildings are covered in Rule 234C. Rule 232A does not apply to building sites under construction. The U.S. Labor Department (OSHA), especially 29CFR1926, as well as most states, have regulations regarding the clearances between cranes and energized power lines.

| *Figure 4-3A* **Mapping the Danger Zone** |

Map and Barricade the 30 foot wide Danger Zone
(15 feet on each side of the powerline poles).

ALWAYS notify the power company before you begin crane
operations near powerlines.

| *Figure 4-3B* Safe Crane Location (Side View) |

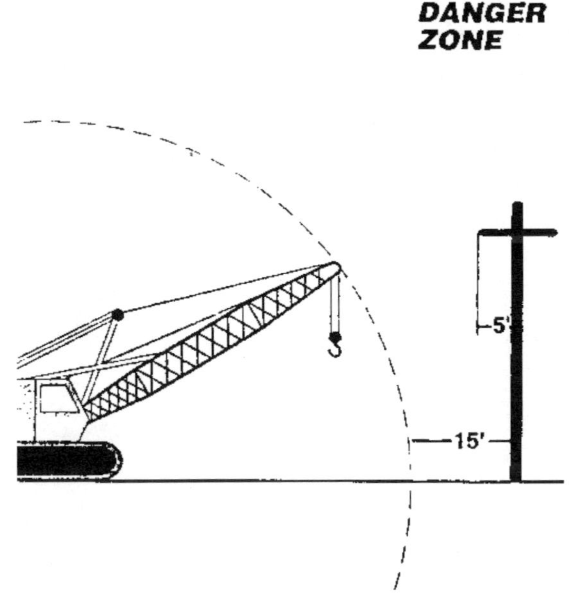

Crane boom is lowered level to the ground. Crane, boom, and boom tip cannot intrude into the danger zone extending 15 feet on each side of the powerline.

| *Figure 4-3C* Plan View of Mapping the Danger Zone |

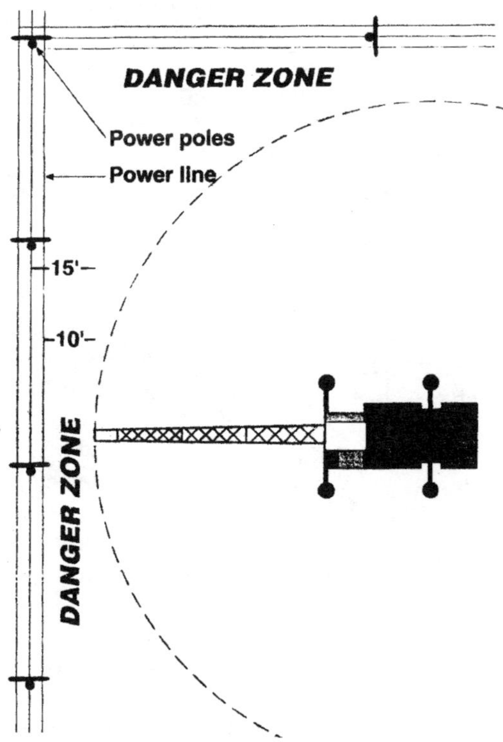

13,800-volt powerline. Alleged lack of safety device on crane such as insulated link and proximity alarm. Work area compromised powerline danger zone. Verdict for the injured.

Plaintiff Brown; occurred November 4, 1981; Delaware County Superior Court, Indiana, #2S-82/444; Appeals Court, Indiana, 4th District, #18A04-8611-CV-00350; and Supreme Court, Indiana, #18S04-9003-CV-181.

Deceased electrocuted when guiding forms being stacked and hoist line contacted 7,200-volt powerline. Jury verdict for plaintiff. Appeal supported plaintiff's contention that crane was inherently dangerous since no feasible safety devices such as proximity alarm and insulated links were installed, but appeal reversed monetary award. Supreme Court reinstated original jury award for plaintiff.

Plaintiff Flatt; Occurred June 8, 1982; U.S. District, Western District of Washington at Tacoma, #83-654T.

Female plaintiff lost leg and most of other foot when crane operator swung mobile crane into 115,000-volt powerline. Alleged there was no organized contractor safety program required by U.S. Department of Energy (Bonneville Power Administration) that supervised work and the crane had no proximity alarm or adequate warning label. Case settled.

Because of the repeated occurrence of crane/powerline contacts, the American Trial Lawyers Association (ATLA), formed a committee for the express purpose of exchanging and discussing information related to crane litigation. (This committee is chaired by Robert J. Mongeluzzi, Daniels, Saltz, Mongeluzzi & Barrett, Ltd., One Liberty Place, 34th Floor, 1650 Market Street, Philadelphia, PA 19103, [215] 496-8282.) The activities of some groups who are involved in safety sometimes do not show as much concern about the continuing death and injury rate from this hazard or, for that matter, any other hazard, as the trial lawyers do. The safety profession should be the first to develop a volunteer means of instituting safety measures to control this hazard. In 1969 the National Safety Council's Utility Section published a summary of such occurrences, but it has never been re-published with current information. In 1981 the National Electric Safety Code, ANSI C2, deleted Sections 210 and 211 which stated:

210. *Design and Construction.* All electric supply and communication lines and equipment shall be of suitable design and construction for the service and conditions under which they are to be operated.

211. *Installation and Maintenance.* All electric supply and communication lines and equipment shall be installed and maintained so as to reduce the hazards to life as far as is practical.

My research suggests that the initial response of some corporate management to litigation is to attempt to conceal hazard data and discredit suggested hazard prevention measures. In view of the estimated $200 million that has been paid for powerline contact claims, a modest investment in hazard prevention programs to control this hazard would not only be very humanitarian, but highly profitable.

Powerline Contact by Aerial Lifts

Definition
Because masts and booms on aerial lifts are often very long, they can easily reach powerlines. There are two types of aerial lifts, those that have non-conductive booms and baskets used by electric utility workers and those that have a conductive boom used by non-utility workers.

Description
Either a phase-to-ground or phase-to-phase electrical contact can be made by those working in an aerial lift if the mast or boom is raised into a bare, uninsulated, high-voltage powerline.

Risks Presented by Hazard
On some vehicle-mounted elevating and rotating work platforms used by electric utility workers, the lower boom section is sometimes uninsulated. If this section makes unintended contact with an energized powerline, anyone touching the vehicle can be electrocuted. Even

Figure 4-4 **Insulated Link**

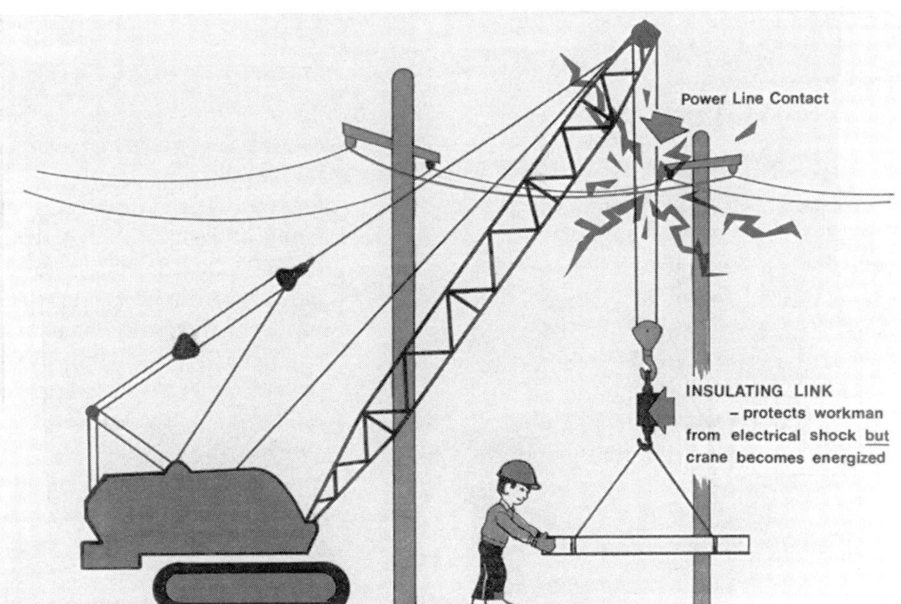

Crane Equipped with an Insulated Link

though the upper boom on some of these may be non-conductive, the attachment system for the basket sometimes has exposed metal bonded to the control lever. If this exposed metal makes unintended contact with an energized powerline while the line person is working on another line, the line person can be electrocuted by phase-to-phase contact if the control lever is touched.

Most aerial lifts used by non-electric utility workers have conductive masts, booms, powered ladders, or other elevating devices. Many workers, when raising or maneuvering this type of lift, have been injured or killed when the lift came into contact with an energized powerline.

Available Hazard Prevention Measures

When electric utilities are buying lifts for line people, they should select those that have both the upper and lower booms insulated to prevent injury to workers who may touch the vehicle. Many manufacturers of this type of lift have developed retrofit kits to prevent phase-to-phase injury to line people working from the basket.

To prevent unintended electrocution when using a lift with a non-conductive boom, a non-conductive barrier should be installed as discussed in Chapter 10, Controls. (Also see Chapter 4, Powerline Contact, for other methods of controlling this hazard.)

OSHA Requirements

1. OSHA 1926.950 through 1926.960 covers electric utility powerline construction.
2. OSHA 1926.550(a)(15).

ANSI Requirements

1. ANSI A92.2, Vehicle-Mounted Elevating and Rotating Aerial Devices.
2. ANSI A92.5, Boom-Supported Elevating Work Platforms, 13.2.2 (1).
3. ANSI A92.6, Self-Propelled Elevating Work Platforms, 14.2.2.

Other References

See Chapter 4, Powerline Contact, on mapping the danger zone.

Suggested Design Criteria

On lifts with conductive booms as used by non-utility workers, provide a non-conductive boom or cage around the aerial basket or platform to serve as a barrier to shield the operator or passengers from direct contact with powerlines. (See Chapter 10, Controls.)

Include detailed mapping instructions in the operator's manual and post point-of-operation warning on the lift. (See previous section: Powerline Contact by Hoist or Boom.)

Representative Litigation

Plaintiff Weldon; occurred June 17, 1974; Superior Court, King County, Seattle, Washington, #789117.

Injured lineman lost arm when touching bare metal control handle bonded to a small metal work boom on an aerial basket when the boom touched an energized powerline while injured was working on an adjoining line. Alleged failure to provide insulated control handle and a non-conductive work boom. Verdict for plaintiff.

Plaintiff Ellet; occurred May 19, 1976; Circuit Court, Cook County, Law Division, Illinois, #77L-337.

Injured volunteer fireman was removing flag from top of flagpole using a metal fire snorkel ladder truck. As he raised the boom to reach the flag, his head touched a powerline and he sustained horrible injuries. Alleged that the snorkel boom should have a protected cage for the operator. Case settled.

Plaintiff Canfield; occurred April 30, 1987; U.S. Northern District, Eastern Division, Ohio, #5:89CV0775.

Deceased lineman was electrocuted when uninsulated jib cylinder angle bracket, located underneath the work platform from which he was stringing line and which was out of his view, contacted a high-voltage powerline as he was moving his platform getting ready to place a de-energized line on a crossarm of a pole. Alleged that the manufacturer placed an uninsulated jib cylinder angle bracket in a dangerous location outside the view of the operator in the work platform. The jib cylinder had originally been located on the upper side of the boom where it was clearly visible. Case settled.

5

Upset

Upset Due to Overloading

Definition

Upset can occur when the tipping load and rated capacity of a crane is exceeded while attempting to lift or maneuver a load.

Description

Cranes can easily upset from overloading. The margin of safety between the actual tipping load and rated capacity often varies from fifteen to twenty-five percent, based on the type of crane. (See Figure 5-1.)

On some types of cranes, the location of the boom in relation to where it is mounted on the carrier determines the margin of safety between tipping load and rated capacity. For a crane mounted on a flatbed truck with the hydraulic boom located directly behind the truck cab, the over-the-rear or over-the-cab tipping load can be as much as twice the rated capacity, and the crane can easily upset when the load is slued (rotated) to either side.

On some high-reach hydraulic cranes, the weight of the boom itself — without a load — can create an imbalance and cause upset when the boom is positioned at a low angle. On some models this can occur even with outriggers extended.

Today's crane operator is confronted with a number of variables that affect lifting capacity:

1. The ability to extend the hydraulic boom while raising or lowering it increases the radius and very swiftly reduces lifting capacity.
2. The choice of extending or retracting outriggers.
3. The relationship between where the boom is positioned at the various points of the compass or clock and its particular carrier frame.
4. The actual weight of the load. Often the operator relies upon his perception, instinct, or experience to judge load weight and may not respond fast enough when the crane begins to feel light.
5. Some overloads result from a load that is not freely suspended, such as the use of a vibratory pile driver.
6. A rough-terrain mobile hydraulic crane, which is designed with a carrier frame with four-wheel drive and steer controlled from the rotating cab, has a great variance in lifting capacity from front to rear. Some carrier frame designs locate the engine, transmission, and hydraulic pump in the rear and the fuel tank in the front, making the crane's tipping load over the front much greater than its tipping load over the rear. A test lift over the front is not valid for a lift over the rear. For this reason, this type of crane should be equipped

with a load moment indicator (LMI) and the operator should be qualified in the use of this device. Further, the load chart and operator's manual should clearly explain this hazard.

7. Only freely suspended loads should be lifted. Freeing a form panel stuck to freshly cured concrete, raising piling with a vibratory pile driver, or lifting a floating log of unknown weight are all loads that are not freely suspended.

8. Sometimes an operator needs to be able to let the load free fall to overcome an unintended overload from loss of stability (such as outrigger failure). Another hazard is the use of logging tongs to guide a heavy floating object. In one situation, management provided the crane operator with logging tongs attached to the hoist line to guide a heavy floating log over the spillway of a dam. When the log was caught by the current and could not be released from the tongs, it pulled the crane into the water and the operator drowned.

9. Flatbed trucks that have hydraulic cranes mounted on them have a very high center of gravity and are known to upset while being driven (roaded) or being parked by the edge of a road where the road shoulder slopes away. It appears that those who install cranes on flatbed trucks only comply with the highway load displacement requirements of the Department of Transportation (DOT) to prevent abuse of the pavement. Little information is provided to the owner or driver of this type of vehicle as to the center of gravity, nor is tilt board test data made available. The operator needs to know at what angle the crane loses stability and can upset. Travel on curves can be very hazardous, as upset can easily occur on a vehicle with a high center of gravity.

All of these variables create error-provocative circumstances for operators that lead to their inadvertently exceeding the tipping load and rated capacity, which can cause the crane to upset or fail structurally.

Risks Presented by Hazard

From review of over one thousand crane upsets occurring over a thirty-year period, I have projected that an upset occurs in about every 10,000 hours of crane use. Nearly 75% of these upsets were the result of error-provocative circumstances that caused the operator to inadvertently exceed the crane's lifting capacity. The good news is that in the years 1993-2003, the occurrence of crane upset from overloading in the categories of outrigger retracted and extended appears to be declining. This trend can be attributed to the industry acceptance and use of LMIs. However, when averages of failure are examined, they remain constant. Even with fewer upsets occurring, the failure modes remain the same, as the older cranes are not equipped with LMIs. The following breakdowns were made:

15% were in the travel mode

39% were making swing with outriggers retracted

15% were making a pick with outriggers retracted

14% were making a pick or swing with outriggers extended

6% were making a pick or swing, use of outriggers unknown

7% were due to outrigger failure

4% were from other activity

3% resulted in fatalities

8% resulted in lost-time injuries

20% resulted in significant damage to property other than the crane

A new hazard is present in some of the newer telescoping hydraulic crane booms. The tipping load may actually be more than 25% above the rated capacity. When the structural strength of the boom is not more than 25%, it is then not strong enough to support a load that is less than the actual tipping load. In this circumstance the operator may feel the crane is stable and not "light" and assume the crane is not overloaded when the boom suddenly collapses because he has exceeded the rated capacity. In addition, a crane can be overloaded beyond its rated capacity if it is not level. When the crane is tilted towards the boom tip, the radius is increased and the rated capacity reduced.

Available Hazard Prevention Measures

During the last thirty years, there has been a great leap forward in available systems to prevent crane upset from overloading. Crane operation is no longer a seat-of-the-pants skill. It now requires both planning and training in the use of the latest load-measuring systems and other technology.

With the advent of solid-state micro-processing electronics, load-measuring systems have evolved. Such systems can sense the actual load as related to boom angle and length, warn the operator as rated capacity is approached, and stop further movement. These systems can be designed to automatically prevent the exceeding of the rated capacity at any boom angle, length, or radius. Today most United States crane manufacturers are promoting the sale of user-friendly load-measuring systems to aid the operator in overcoming error-provocative overload circumstances.

Devices such as vibratory pile drivers should only be used with the manufacturer's authorization and a load-sensing device to prevent exceeding the rated capacity. When a crane does not have a level indicator, an operator can sometimes be unaware that the outriggers or crawler treads have sunk into the ground because of poor soil.

For years the only control to avoid upset from overload has been to rely upon an operator's performance and use of load charts. Such charts are often difficult to read. In "Rating Chart for Cantilevered Boom Cranes," SAE J1257 (1979; revised October, 1980), the Society of Automotive Engineers developed criteria designed to improve load chart clarity. Improved charts have helped crane operators understand the lifting capacities of the cranes they use. Formal training should be provided for all crane operators to assure competency in the use of crane load charts and load-measuring systems.

No load should ever be lifted which is not freely suspended. Devices such as vibratory pile drivers should not be used unless authorized by the manufacturer on how it can be safely done.

OSHA Requirements

OSHA has not commented on load-measuring and/or load-moment devices. The Occupational Safety and Health Act of 1970, Section 5(a)(1), General Duties Clause, may be used to address this hazard. OSHA 1910.179(b)(5), 1910.180(c), 1910.181(c), and 1926.550(a)(1) require compliance with the manufacturer's rating charts. OSHA 1910.180(h)(3)(vii) states: "On truck-mounted cranes, no loads shall be lifted over the front area except as approved by the crane manufacturer." Lifting over the front of a truck-mounted crane greatly increases the tipping load, and the boom and turntable may be over-stressed. Sluing of the load to either side would most certainly result in upset. Marine Terminal Standard 1917.46 and Longshoring Standard 1918.74(a)(9) address load-indicating devices. Rating charts and overloading are addressed in 1917.45(b) and 1918.74(a)(2). Counterweight restrictions are noted in 1917.45(f)(6).

ANSI Requirements

ANSI requirements address compliance with manufacturers' rating charts and currently specify the following:

> All new cranes with a maximum rated load capacity of 3 tons or more should have load indicators. (ASME B30.5-1994 *Mobile and Locomotive Cranes*, and its 1995 Addenda B30.5a, paragraph 5-1.9.10, and the same paragraph is in the B30.5, 2000.)

Other References

1. *Safety and Health Requirements Manual*, EM 385-1-1, U.S. Army Corps of Engineers, published in October 1992, Section 16.D.01.a. reads: "All lattice boom and hydraulic mobile cranes shall be equipped with the following: a. A boom angle indicator and a load indicating device, or a load moment indicator (rated capacity indicator): calibration and testing of indicators will be performed in accordance with the manufacturer's recommendations. This requirement is effective January 1, 1994...."

2. Most Western European countries have required load-sensing devices since 1960.

3. The United Kingdom in its 1961 *Factories*, No. 1581, Part III, "Indication of Safe Working Load of Jib Cranes," paragraph 30, requires load-sensing devices for all cranes with a lifting ca-

pacity of over 2,000 pounds.

4. The Netherlands has the most advanced approach to preventing crane upset from overloading by requiring a Load Moment Limiter (LML) system. As specified by the Netherlands Normalization Institute (NNI), all cranes are required to meet NEN norm standards 2017-2028:

   a. The LML insures that the crane cannot lift a load which exceeds its rated capacity for that particular radius.

   b. The LML stops all crane movements leading to an increased or to other hazardous circumstances except sluing.

   c. The crane cannot be started automatically after being switched off by an LML.

   Additional data concerning Dutch crane safety ("Safety the Dutch Way") was published in the March, 1991 issue of *Cranes Today*, p. 31. It is probable that LML will be adopted by all of the European countries that are member states of the Economic Community.

5. *Mobile Crane Manual* by D. E. Dickie of the Canadian Construction Safety Association of Ontario (1982). Chapters 4, 5, 6, and 7 provide excellent discussion on how to use load charts.

6. Numerous U.S. Military Specifications published by the Department of Defense that require load-measuring and load-moment devices on mobile cranes show an acceptance and the necessity for load-measuring devices and are listed as follows:

   a. MIL-C-27840B (USAF), "Crane, Truck Mounted A/S32H-11," Paragraph 3.5.3.4, November 6, 1964.

   b. MIL-C-27718B (USAF), "Crane, Truck Mounted A/S32A-16," Paragraph 3.5.3.4, August 27, 1965.

   c. MIL-C-27840C (USAF), "Crane, Truck Mounted A/S32H-11, 4,000 Crane Capacity, 4x4," Paragraph 3.5.3.4, April 12, 1966.

   d. MIL-T-62089(AT), "Truck, Maintenance; with Rotating Hydraulic Derrick, Air Transportable, 34,500 pounds, GVW, 6x4," Paragraph 3.6.4.5, March 26. 1968,

   e. MIL-C-28564(YD), "Cranes, Hydraulic, Truck Mounted, 8x4, 25-, 40-, and 55-ton Capacities," Paragraph 3.5.4, December 30, 1969.

   f. MIL-C-27718C (USAF), "Crane, Hydraulic, Truck Mounted, A/S32A16, 16,500 Pound Crane Capacity, 6x4, GED," Paragraph 3.5.5.4, September 8, 1970.

   g. MIL-C-28614(YD), "Crane, Hydraulic, Rough-Terrain, 25-ton Capacity, Full Revolving, Wheel-Mounted, 4x4, DED," Paragraph 3.6.7., April 8, 1971.

   h. MIL-C-28622(YD), "Crane, Hydraulic, Wheel Mounted, 4x4, Full Revolving, 45-ton Capacity, DED," Paragraph 3.6.7, September 30, 1971.

   i. MIL-C-28629(YD), "Crane, Hydraulic, Wheel Mounted, 4x4, Full Revolving, (Shipboard Aircraft Crash-Salvage)," Paragraph 3.6.7, February 29, 1972.

   j. MIL-C-52341E(ME), "Crane, Wheel Mounted, Diesel Engine-Driven, 20-ton, Rough-Terrain," Paragraph 3.12.13.7, October 24, 1972.

   k. MIL-T-62089A(AT),"Truck, Maintenance; with Rotating Hydraulic Derrick, Air Transportable, 34,500 pounds GVW, 6x4," Paragraph 3.6.2.5, December 18, 1973.

   l. MIL-C-52326B, "Crane, Crawler Mounted: 60-ton at 25-foot Radius, Diesel-Engine-Driven," Paragraph 3.15.1.4, April 4,1974.

   m. MIL-C-29352(MC), "Crane, Wheel-Mounted: Hydraulic, Rough-Terrain, 30-ton Capacity, Full Revolving, 4x4, DED," Paragraph 3.3.2.9, July 17, 1974.

   n. MIL-C-28614A(YD), "Cranes, Hydraulic, Full Revolving, Wheel Mounted, 4x4, DED," Paragraph 3.10.7, June 12, 1975.

Figure 5-1 **Upset Crane**

o. MIL-C-62253(AT), "Crane: Material Handling (Rebuild)," Paragraph 3.5, September 16, 1975.

p. MIL-C-62135A(AT), "Crane: Material Handling," Paragraph 3.5, May 21, 1976.

q. MIL-C-529911(ME), "Crane, Truck Mounted, Hydraulic, 25-ton, Commercial Construction Equipment, (CCE)," Paragraph 3.4.10, September 27, 1976.

r. MIL-T-62089B(AT), "Truck, Maintenance; with Rotating Hydraulic Derrick, Air Transportable, 34,500 pounds GVW, 6x4," Paragraph 3.6.2.5, June 9, 1980.

7. The Society of Automotive Engineers (SAE) first developed a number of standards that relate to load-measuring systems and associated operator aids as published in the following years:

a. "Crane Load Stability Test," SAE J765, October, 1990.

b. "Mobile Crane Stability Ratings," SAE J1289, April, 1981.

c. "Telescoping Boom Length Indicating System," SAE J1180, October, 1980.

d. "Radius-of-Load and Boom Angle Measuring System," SAE J375, April, 1985.

e. "Load Indicating Devices in Lifting Crane Service," SAE J376, April 1985.

f. "Crane Load Moment System," SAE J159, April, 1985.

Suggested Design Criteria

Load-measuring systems that have the capability of interceding and preventing crane operators from exceeding the rated capacity for any boom radius should be included on all cranes.

Retrofit programs should be developed so crane owners can update their equipment with state-of-the-art load-measuring systems.

Patents for Jib Load Limiting Devices

A load limiting device for a jib extension includes: a pivotal insert adjacent to the tip of the jib extension; a cable sheave mounted on the insert by a load on the cable; an angle transducer for measuring the angle of orientation; and the means for determining, based upon such force and angle of orientation, the load on the jib extension.

U.S. Patents

DATE	NUMBER	INVENTOR
Feb. 1936	2030529	Nash
Aug. 1966	3266638	Popov

DATE	NUMBER	INVENTOR
Feb. 1972	3638781	Comley
Aug. 1974	3827514	Bradley
Sep. 1974	3833130	Gerdes et al.
Dec. 1975	3924752	Hoofnagle
Nov. 1976	3990584	Strawson et al.
Jan. 1977	4003482	Cheze
Nov. 1977	4057792	Pietzsch et al.
Jan. 1979	4133032	Spurling
Nov. 1981	4300134	Paciorek
Jan. 1983	4368824	Thomasson
Nov. 1983	4413691	Wetzel
Apr. 1985	4509376	Thomasson
Jul. 1997	5645181	Ichiba et al.
Jan. 1998	5711440	Wada
Apr. 2000	6044991	Freudenthal et al.
Jan. 2001	6170681	Yoshimatsu

Foreign Patents

Dec. 1962	913507	GB
Apr. 1992	04112190	JP

Representative Litigation

The failure to equip cranes with, or to use, load-measuring systems is almost as frequent and probably as costly as powerline contact. America's crane safety requirements for load-measuring systems are three decades behind those of Western Europe. We are a world leader in electronics, computer technology, smart weapons, and other systems, but we have less than a Third World sophistication when it comes to adopting the latest technology on our cranes. The English magazine, *Cranes Today*, in a news release in its October, 1988, issue, reported that Wylie Weighload, one of Europe's leading manufacturers of load-measuring systems, was undertaking U.S. expansion that "will give the company direct access to the U.S. market at a time when the indigenous crane manufacturers are starting to fit safe load indicators as standard in response to product liability pressures" [emphasis added]. This indicates that the threat of litigation is advancing the application of available safety technology.

Plaintiff Glass; filed 1985; U.S. District, Central Division, Utah, #85-C-550A.

The injured was welding adjacent to craning activity. He was crushed by the boom when crane upset and became a paraplegic. Testimony of English crane safety specialists Ian Aitken, John Cyril Allen, and Peter Oram supported the use of safe-load indicators as required in England. Case settled.

Plaintiff Salmon; occurred September 19, 1979; Superior Court, Riverside County, California, #29429.

Deceased was crushed when guiding a gasoline storage tank into an excavation for a new gas station when the crane upset. Alleged that confined work space required the load radius to be exceeded. Plaintiff contended injury was caused by poor planning and lack of load-measuring system. Case settled.

Plaintiff Baird; occurred October 20, 1980; U.S. District Court, District of New Jersey, #81-3905.

Injured struck by crane boom when the crane tipped over when reaching for the loading hook to pick a light load. Crane upset because of weight of boom. Plaintiff contended injury was caused by two hazards: lack of load- measuring device and too small outrigger pads that may have sunk into ground. Case settled.

Failure to Use Outriggers/Soft Ground and Structural Failure

Definition

Crane upset can occur when an operator does not extend the outriggers, when a crane is positioned on soft ground, and when the outriggers are inadequate.

Description

Many cranes upset because use of outriggers is voluntary and usually left to the discretion of the operator who might not perceive a potential hazard. In some circumstances the job site foreman orders the crane operator to use the crane with the outriggers retracted. Sometimes an operator cannot extend the outriggers because of insufficient space or a work circumstance that arises when pre-job planning is not done. Often, outrigger pads are too small to support the crane on hard ground, let alone on soft ground, which poses a problem in itself.

A mobile hydraulic rough-terrain crane on rubber with the outriggers retracted is completely unstable on a side lift. Whether a crane is on rubber or truck-mounted, if outriggers are not extended, the lifting capacity at side angles drops dramatically. As the crane boom rotates on the turntable, upset occurs so quickly that the operator cannot perceive the loss of stability until too late. Sometimes the weight of the boom alone at a low angle will upset the crane even with outriggers properly extended.

In a few instances, outriggers have collapsed because they were not strong enough or had been damaged.

The rear outrigger on some cranes can disengage from the float or pad when the load is first lifted. When the crane cab/boom is swung around 180 degrees, the outrigger can slip from such an unfixed pad connection and cause the crane to upset or the boom to buckle as a result of loss of stability.

A small truck-mounted crane can be upset due to foreseeable human error when outriggers cannot be extended or lowered on one side to work in a confined location or are not repositioned after having been momentarily retracted or raised to let traffic by. (See previous section: Upset Due to Overloading.)

Upset due to soil failure occurs because either the ground is too soft or the outrigger pads are not big enough. Soil conditions range from wet sand that can only support 2,000 pounds per square foot, to dry hard clay that can support 4,000 pounds per square foot, to a well-cemented hardpan that can support as much as 10,000 pounds per square foot.

Risks Presented by Hazard

As stated earlier in this chapter, an analysis of some 1,000 crane upsets over a twenty-year period showed that over half of the incidents involving this hazard occurred when the crane operator was either swinging the cab or extending or lowering a telescoping boom without outriggers extended. These actions rapidly increase the lifting radius and upset occurs quickly.

D. E. Dickie of the Canadian Construction Safety Association of Ontario states in *Crane Handbook* that fifty percent of mobile crane failures are caused by improper use of outriggers.

Available Hazard Prevention Measures

Since such a high proportion of upsets occur when outriggers are not extended, I believe that design changes to overcome this hazard are needed. When human failure is as predictable as this, the surest way to avoid upset is to make the machine inoperable until the operator extends or lowers the outriggers. Some aerial basket designs include limit switches to prevent boom movement until outriggers are extended and in place to avert upset. The newer aerial basket trucks have hydraulic systems with interlocks that preclude boom operation until outriggers are fully extended and fully supporting the machine, with wheels completely off the ground. Interlocks are required in Military Specification MIL-T-62089(AT), "Truck, Maintenance; with Rotating Hydraulic Derrick, Air Transportable, 34,500 Pounds, GVW, 6x4," dated March 26, 1968, and its updates, MIL-T-62089(A), December 18, 1973, and MIL-T-62089(B), June 9, 1980.

When poor soil is encountered or when outriggers have inadequate floats or pads, well-designed blocking or cribbing is needed under the outriggers. On all types of cranes, when floats are used, ANSI requires that they be securely attached. It also requires that blocking used to support outriggers shall be strong enough to prevent crushing, be free of defects, and be of sufficient width and length to prevent shifting or toppling under load.

OSHA Requirements

OSHA 1910.180(h)(3)(ix), Marine Terminal Standard 1917.45(f)(7), and Longshoring Standard 1918.74(a)(2) discusses outriggers.

ANSI Requirements

ANSI B30.5a, *Mobile and Locomotive Cranes*, and Addenda 2000, Paragraph 5-1.9.3, and ANSI B30.15, *Mobile Hydraulic Cranes*, require the use of outriggers for the various types of mobile cranes.

Other References

1. *Safety and Health Requirements Manual*, EM 385-1-1 (U.S. Army Corps of Engineers, November 3, 2003), Sections 16.D.09 read:

   16.D.09 Outriggers.

   a. When the load is to handled and/or

the operating radius require the use of outriggers, or any time when outriggers are used, outriggers shall be fully extended and set to remove the machine weight from wheels (except locomotive cranes).

b. When outrigger floats are used they shall be securely attached to the outriggers.

c. Blocking under outriggers floats shall meet the following requirements:
   (1) sufficient strength to prevent crushing, bending, or shear failure,
   (2) such thickness, width, and length as to completely support the float, transmit the load to the supporting surface, and prevent shifting, toppling, or excessive settlement under load, and
   (3) use of blocking only under the outer bearing surface of the extended outrigger beam floats.

16.D.10.

Unless the manufacturer has specified an on-rubber rating, mobile cranes shall not pick or swing loads over the side of the crane unless the outriggers (if so equipped) are down and fully extended.

2. *Crane Handbook* by D. E. Dickie of the Canadian Construction Safety Association of Ontario (1972) pp. 69 and 75, Figures 3.42-3.49.

Suggested Design Criteria

Designers should assure that outrigger pads are of sufficient size to afford support on wet sand and/or provide sufficient information in the operating manual on how to provide sufficient support.

If the load-measuring systems on rough-terrain cranes designed for on-rubber and off-rubber use do not function during on-rubber lifts, manufacturers of such cranes should provide interlocks that prevent boom extension and sluing over-the-side beyond the wheels. (See Figure 5-2.)

Representative Litigation

Plaintiff Baird; occurred October 20, 1980; U.S. District Court, District of New Jersey, #81-3905.

Injured struck by crane boom when the crane tipped over when reaching for the loading hook to pick a light load. Crane upset because of weight of boom. Plaintiff contended upset was caused by two hazards: lack of load- measuring device and too-small outrigger pads that may have sunk into ground. Case settled.

Plaintiff Garcia; occurred 1978; Harris County, Texas, #1,153,990.

Deceased struck by the boom of a crane when it upset with outriggers not fully extended. Crane rental firm alleged crane manufacturer should have installed limit switches to restrict sluing of crane boom when outriggers are not in place and load-measuring system. Case settled.

Plaintiff Pritts; occurred December 19, 1974; Court of Common Pleas, Allegheny County, Pennsylvania, #G.D.75-25480.

Two men injured while in a skip adjusting concrete forms on bridge pier under construction. Outriggers not fully extended. It was alleged that limit switches were necessary to prevent boom from sluing when outriggers were not in place. Case settled.

Plaintiff Humm; occurred February 27, 1976; Circuit Court, 3rd Judicial Circuit, Madison County, Illinois, #76-L-482.

Injured in skip being lowered when crane upset with outriggers retracted and boom was sluing. It was alleged that limit switches were needed to prevent boom sluing when outriggers were retracted or load indicators used. Case settled.

Plaintiff Renkin; occurred April, 1981; Superior Court, Snohomish County, Washington, #83-02-02744-4.

Injured sustained paraplegia unloading concrete pipe. Boom was partially extended. Outrigger pad sank into ground and he began to boom down. Bolts broke and boom disconnected from pedestal, striking injured. Case settled.

| *Figure 5-2* **Interlocks Limit Boom Movement** |

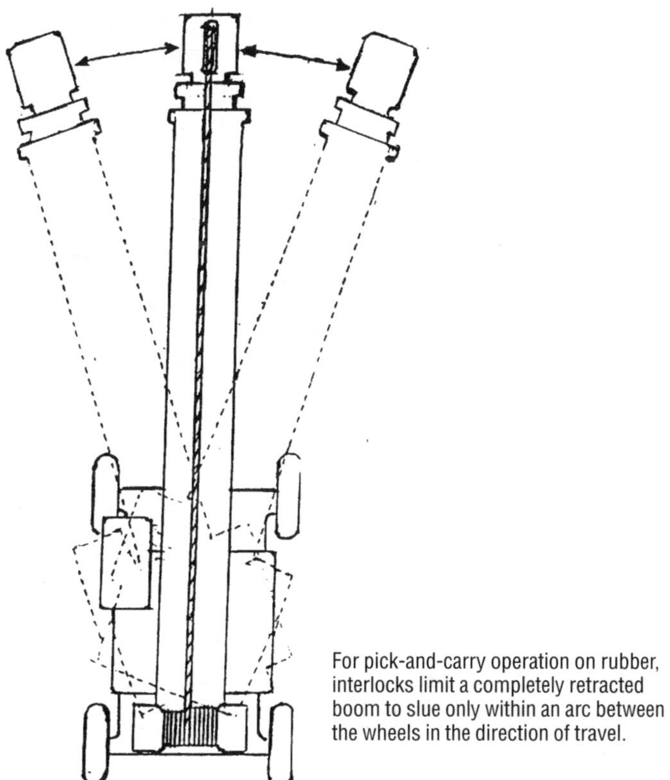

For pick-and-carry operation on rubber, interlocks limit a completely retracted boom to slue only within an arc between the wheels in the direction of travel.

Lift Simulation

Computer Assisted Design (CAD) software is being used to develop lift planning. Multiple-crane lifting in the past has always been a *tongue in cheek* process with success being almost a matter of chance. The first step in the use of this technology is to develop a computer profile of each crane's lifting capacity in all configurations. Next is to develop a computer library of the various rigging configurations, including lifting beams and other accessories. The third is to develop three-dimensional analysis of the proposed lift that is well within the safe design-rated capacities of the various crane and rigging configurations for the planned lift. It is with this planning that difficult lifts can be made within the rated capacities of the various cranes, avoiding two-blocking and limiting the boom extensions and angles. This planning quickly eliminates cranes whose lifting profile is not within safe limits for the proposed lift. A number of software products are available that

can be adopted for such programs, but be sure that you retain the expertise of someone familiar with cranes and their lifting capacities who has previously written these highly specialized programs.

The greatest value in the use of this lift planning technology is that a number of dry runs can be made where crane operators, rigging crews, and project management can view each step of the proposed critical lift on the computer screen. As cranes are designed with greater lifting capacity, so will the assembly of larger and larger structural and process components be a common objective. As these methods of construction evolve, the computer planner needs to include basic data, such as soil conditions supporting the outriggers and the capabilities of shackles, cables, pins, and rigging accessories. As this technology advances and becomes refined, it can provide excellent guidance in making everyday lifts free from error and save lives.

Aerial Lift Upset When Outriggers Not Extended

Definition
When outriggers are not extended and in place, any aerial lift or elevated work platform can upset.

Description
Operators are often unaware of the propensity of aerial lifts or work platforms to upset when outriggers are not extended and in place. Some lifts allow limited boom movement without upsetting. For this reason operators may not always extend outriggers. (See Figure 5-3.)

Some lifts allow the operator to make a nearly vertical lift without upsetting when outriggers are not extended. However, any movement away from the nearly vertical position can cause upset.

Risks Presented by Hazard
Outriggers systems that rely upon an operator's judgment on whether to extend or not extend outriggers are error-provocative. A lift that allows the operator to choose whether to extend or not extend outriggers can create circumstances that lead to upset and the operator can fall many feet, depending upon the location of the basket or work platform above the ground.

Available Hazard Prevention Measures
Ask the manufacturer to install interlocks that prevent movement of the boom when outriggers are not extended and in place.

OSHA Requirements
OSHA refers to ANSI A92.2-2001, *Vehicle-Mounted Elevating and Rotating Aerial Devices*, but this ANSI standard has no requirements for interlocks.

ANSI Requirements
1. *Boom-Supported Elevating Work Platforms*, ANSI A92.5, Outriggers, Stabilizers, and Extendible Axles, states: "Where the work platform is equipped with outriggers, stabilizers, or extendible axles, interlocks should be provided to ensure that the platform cannot be positioned beyond the maximum travel height unless the outriggers, stabilizers, or extendible axles are properly set. Control circuits shall ensure that the driving motor(s) cannot be activated unless the outriggers or stabilizers are disengaged and the platform has been lowered to the maximum travel height (MTH)."

Figure 5-3 Aerial Basket Upset with Outriggers Retracted

2. *Self-Propelled Elevating Work Platforms,* ANSI A92.6-1979, section 4.2 states: "Machines shall use interlock means that will prevent driving the unit unless the platform height, platform configuration, or any combination of the foregoing, are adjusted to meet the stability-test requirements...." Section 4.4 states: "Where outriggers, stabilizers, or extendible axles are required to meet the side load test described in 4.3, interlocks shall prevent the platform from being raised above the height at which these devices are required unless the required devices are extended. Interlocks shall also prevent the retraction of these devices while the platform is above that level."

Other References

Bibliographic references 125-133 on system safety provide helpful guidance on how to conduct a fault-tree analysis that would reject error-provocative design.

Suggested Design Criteria

When the stability of an aerial lift or work platform relies upon the extension of outriggers, interlocks should be provided as standard equipment so the user is unable to raise the lift or work platform until the outriggers are extended and in place.

Representative Litigation

Plaintiff Whalen; occurred June 12, 1979; 3rd Judicial Circuit; Madison County, Illinois, #81-L-194.

Contractor's employee was injured when he fell with upsetting aerial work platform that did not have outriggers extended. No interlocks to prevent boom movement until outriggers were extended and no load-measuring device to warn operator that he was exceeding rated capacity. Alleged that manufacturer's retrofit program had failed to notify the platform owner of the available interlock retrofit kit so the equipment remained in use without available safeguards. Owner also allowed untrained contractor's employees to use equipment. Case settled during trial.

Plaintiff Rupe; occurred January 16, 1985; Federal District Court, Kansas City, Kansas, #862488-O.

Injured fell sixty feet when aerial work platform upset. Alleged no interlocks to prevent boom movement when outriggers not extended. Case settled during trial.

6

Two-Blocking

Definition

The Dictionary of Terms Used in the Safety Profession, Fourth Edition (American Society of Safety Engineers (ASSE), 2001), defines two-blocking as: "In crane operation, the condition in which the lower load block (or hook assembly) comes in contact with the upper load block (or boom point sheave assembly), seriously interfering with safe operation of the crane."

Description

Both latticework- and hydraulic-boom cranes are prone to two-blocking. On latticework booms, the following stresses in combination can break the hoist line:

1. the weight the load
2. the weight of the headache ball and rigging
3. the tension cause by the hoist drum
4. the weight of the boom when pendant guys go slack
5. whip caused by a long latticework boom

The power of the hydraulic rams that extend hydraulic booms is often sufficient to break the hoist line if the line two-blocks. Both latticework and hydraulic-boom cranes will two-block when the hook is near the tip and the boom is lowered. An operator can forget to release or pay out the load line when extending the boom and the hoist line can be broken.

A crane equipped with a jib or a flyline on the boom is also prone to two-blocks. An operator can easily be unaware of a two-block-ing situation that may be arising on the line not being used. Human factors logic tells us that when an operator must use two controls, one for the hoist line and one for the hydraulic boom extension, the chance of error is increased.

Risks Presented by Hazard

The ASSE *Dictionary of Terms Used in the Safety Profession* also states: "When two-blocking occurs, life-threatening forces can be applied to the hoist or hook arrangement, either breaking the hoist line or disengaging the load straps from the hook and causing the load hook to fall or lose the load, imperiling the lives of those working or standing directly below."

Death and crippling injuries often result when a line two-blocks. If the load line breaks, and it happens to be supporting a worker on a bosun's chair, several workers on a floating scaffold, or a load above people, catastrophe can result. Over the years there have probably been thousands of two-blocking occurrences that have broken the hoist line, with most of them going unrecorded because no one was injured when the hoist line failed and dropped the hook and/or load.

Available Hazard Prevention Measures

Anti-two-blocking devices have been available for years, but industry acceptance has lagged when it comes to adopting them as preventive measures. There are several ways to prevent two-blocking:

1. An electrical sensing device. A weighted ring around the hoist line can

be suspended on a chain from a limit switch attached to the boom tip. When the hoist block or headache ball touches the suspended, weighted ring, the limit switch opens and an alarm warns the operator. It can also be wired to intercede and stop the hoisting. The circuitry is no more complex than an electric door bell. (See Figure 6-1.)

2. On hydraulic cranes the hydraulic valving can be sequenced to pay out the hoist line when the boom is being extended, thus avoiding two-blocking.

3. When making a lift, sufficient boom length is necessary to accommodate both the boom angle and sufficient space for rigging, such as slings, spreader bars, and straps. To avoid bringing the hook and headache ball into contact with the boom tip, a boom length of 150 percent of the height of the intended lift is required for a boom angle of 45 degrees or more.

Anti-two-blocking devices should be standard equipment on all cranes. Currently, most new mobile hydraulic cranes are being equipped with anti-two-blocking systems by the manufacturer.

OSHA Requirements

1. OSHA 1926.550(g)(3)(ii)(c) only requires an anti-two-blocking device on cranes being used to suspend a personnel platform.

2. It is interesting to note that OSHA 1910.179(g)(5)(iv) for overhead and gantry cranes requires: "The hoisting motion of all electric traveling cranes shall be provided with an overtravel limit switch in the hoisting direction." This is, in itself, an anti-two-blocking system.

3. OSHA 1910.179(n)(4)(i), "Hoist Limit Switch," requires: "At the beginning of each operator's shift, the upper limit switch of each hoist shall be tried out under no load...." This requirement shows that anti-two-blocking concepts have been applied on overhead and gantry cranes.

4. OSHA Subpart N, 1926.550(a)(1) requires: "The employer shall comply with the manufacturer's specifications and limitations applicable to the operation of any and all cranes and derricks." In some instances, crane owners have failed to provide functioning anti-two-blocking devices as recommended by the manufacturers.

ANSI Requirements

1. The old ANSI B30.15-1975, *Mobile Hydraulic Cranes*, Section 15-1.3.2(d), states: "On a telescoping-boom crane, with less than 60 feet of extended boom, a two-blocking damage preventive feature shall be provided capable of preventing damage to the hoist rope, and/or other machine components, when hoisting the load, extending the boom, or lowering a boom on a machine having a stationary winch mounted to the rear of the boom hinge." Considering the fact that the longer the boom, the further the boom tip is from the observing operator, it seems to me that an anti-two-blocking device requirement limited to booms less than sixty feet is insufficient.

2. ANSI B30.5-1989 recommends use of an anti-two-blocking device solely as a damage-preventive feature for the boom tip. In my opinion, this standard seems to fail to identify two-blocking as a life-threatening hazard. Because of the dire consequences associated with this hazard, from a human factors standpoint, such devices should be standard equipment on all cranes. The December 1992 Addendum now requires:

   5-1.9.9 Two-Blocking Features

   (a) Telescopic boom cranes manufactured after February 28, 1992 shall be equipped with an anti-two-block device or a two-block damage prevention feature for all points of two-blocking. Telescopic boom cranes manufactured before February 28, 1992 should be equipped with a two-block warning feature(s), a two-block damage prevention feature, or an anti-two-block device for all points of two-blocking (i.e., jibs, extensions, etc.).

(b) Lattice boom cranes manufactured after February 28, 1992 shall be equipped with a two-block warning feature which functions for all points of two-blocking. Lattice boom cranes manufactured before February 28, 1992 should be equipped with a two-block warning feature which functions for all points of two-blocking.

(c) The current standard ASME B30.5a, Addenda 1995, paragraph 5-1.9.9 (a)&(b) require the same as (b) above.

3. ANSI B30.5 2000, Section 5-1.9.9 Two-Blocking features

(a) Telescopic boom cranes shall be equipped with an anti-two-block device or a two-block damage prevention feature for all points of two-blocking (i.e. jibs, extensions, etc.). (See Section IV, New and Existing Installations).

(b) Lattice boom cranes shall be equipped with an anti-two-block device or a two-block warning feature that functions for all points of two-blocking (See Section IV).

4. ANSI B30.5 Section IV- New and Existing Equipment

(a) *Effective Date.* The effective date of this volume for the purpose of defining new and existing installations shall be 1 year after its date of issuance.

(b) *New Installations.* Construction, installation, inspection, testing, maintenance, and operation of equipment manufactured and facilities constructed after the effective date of this volume shall conform with the mandatory requirements of this volume.

(c) *Existing Installations.* Inspection, testing, maintenance, and operation of equipment manufactured and facilities constructed prior to the effective date of this volume shall be done, as applicable, in accordance with the requirements of this volume. It is not the intent of this volume to require retrofitting of existing equipment. However, when an item is being modified, its performance requirement shall be reviewed relative to the current volume. If the performance differs substantially, the need to meet the current requirement shall be evaluated by a qualified person selected by the owner (user). Recommended changes shall be made by the owner (user) within one year.

Other References

1. Society of Automotive Engineers (SAE) standard "Two-Block Warning and Limit Systems in Lifting Crane Service," J1305, June, 1987, provides a recommended practice for minimum performance requirement of devices that signal or automatically prevent two-blocking.

2. Military Specifications have been published by the Department of Defense that require anti-two-blocking devices as follows:

a. MIL-C-27718B (USAF), "Crane, Truck-Mounted, A/S32A-16," Paragraph 3.7.8, August 27, 1965.

b. MIL-C-27840C (USAF), "Crane, Truck-Mounted, A/S32H11, 4,000 lb. Capacity, 4x4," Paragraph 3.7.6, April 12, 1966.

c. MIL-C-27718C (USAF), "Crane, Hydraulic, Truck-Mounted, AS32A-16, 16,500 lb. Capacity, 6x4, GED," Paragraph 3.7.8, September 8, 1970.

d. MIL-C-28616(YD), "Crane, Shovel, Convertible, Crawler-Mounted, Commercial DED," Paragraph 3.9.1, April 30, 1971.

e. MIL-C-28622(YD), "Crane, Hydraulic, Wheel-Mounted, 4x4, Full Revolving (Shipboard Aircraft Crash Salvage)," Paragraph 3.6.2, September 30, 1971.

f. MIL-C-23877B (YD), "Cranes, Hydraulic, Wheel-Mounted, 5 & 12 ½ Ton Capacities," Paragraph 3.6.4, May 1, 1972.

g. MIL-C-28614A(YD), "Crane, Hydraulic, Full-Revolving, Wheel Mounted, 4x4, DED," Paragraph 3.10.2, June 12, 1975.

3. Holland, Belgium, France, Germany, Finland, Sweden, Switzerland, Czechoslovakia, the Soviet Union, and Japan require anti-two-blocking equipment on cranes. The United Kingdom (Great Britain) has no specific requirement for an anti-two-blocking device but is able to achieve the same end results by limiting the hoisting capacity to the rated capacity. Thereby no load is imposed upon a hoist cable that exceeds the rated capacity, which is well-below the breaking strength of the hoist line. In short, all European countries have arrived at methods to prevent the two-blocking failure mode.

4. *Crane Handbook* by D. E. Dickie of the Canadian Construction Safety Association of Ontario (1972), p. 130, Figure 4.62, recognizes that the hazard of two-blocking will cause the hoist line to break.

5. The states of Oregon, California, Utah, and Minnesota have anti-two-blocking requirements.

6. *Safety and Health Requirements Manual,* EM 385-1-1 (U.S. Army Corps of Engineers, October, 1992), Section 16.D.01.e. requires anti-two-blocking devices as follows:

(1) Lattice boom cranes shall be equipped with an anti-two block device to stop the load hoisting function before the load block or load contacts the boom tip....

(2) Telescopic boom cranes shall be equipped with an anti-two block device to stop the load hoisting function before the load block or load contacts the boom tip and to prevent damage to the hoist rope or other machine components when extending the boom.

(3) Telescopic boom cranes which are used exclusively for duty cycle operations (such as clamshell, dragline, grapple, or pile driving) shall be equipped with a two-blocking damage prevention feature or warning device to prevent damage to the hoist rope or other machine components when extending the boom.

7. The following patents directly relate to anti-two-blocking devices, or by their nature avoid the circumstances that lead to two-blocking, and show that two-blocking circumstances are so hazardous that numerous patents have been filed:

Patents for Anti-Two Blocking Devices

DATE	NUMBER	INVENTOR	ITEM
07.24.06	826,529	Baldwin & Ihlder	Automatic Stop for Whip Hoists
04.11.22	1,412,643	Franklin E. Arndt	Safety Stop for Hoisting Drums
09.02.24	1,506,757	E. S. Lammers	Control System
08.26.30	1,774,440	L. P. Hutt	Limit Switch
09.09.30	1,775,435	E. H. Lichtenberg	Boom Hoist Safety Means
12.08.31	1,834,985	F. Stoner	Industrial Truck
05.10.32	1,857,172	C. H. Wagner	Safety Mechanism for Material Machines
07.12.32	1,867,452	H. E. Hallenbeck	Combined Indicator and Limit Switch
09.13.32	1,877,171	H. E. Hallenbeck	Switch Control Mechanisms
11.27.34	1,982,339	Ralph Ehrenfeld	Claims Switch
09.10.35	1,013,690	R. C. Lamond	Hoisting Mechanism
02.11.36	2,030,529	Archibald Nash	Safety Device for Cranes
07.01.41	2,248,010	Russell Nelles	Automatic Circuit Controller
12.16.41	2,266,660	Sebastian Sloan	Cable Safeguards

DATE	NUMBER	INVENTOR	ITEM
05.18.43	2,319,299	Frederick Converse	Means of Measuring Loads
08.31.43	2,328,266	Stanley Dubin	Operating Mechanism for Switches
09.07.43	2,328,697	D. G. White	Limit and Control System Combination
09.21.43	1,330,060	H. P. Kuehni	Safe Load Indicator for Crane Hoists
06.26.51	2,558,517	Philip Handelman	Adjustable Limit Switch Means
12.18.51	2,579,317	J. A. Hepperlen	Control System
12.01.53	2,661,405	James Western	Safety Down Limit Switch for Cable Hoists
11.22.55	2,724,796	E. J. Posselt	Control System for Electric Motor Driven Hoists and the Like
06.24.58	2,840,244	T. W. Thomas	Boom Stop Ram
07.07.59	2,893,134	Reeford P. Shea and Raudenbush	Automatic Leveling Control and Clinometer
02.26.63	3,079,080	H. L. Mason	Crane Warning System
06.18.63	3,094,221	C. W. Galuska	Boom Limit Safety Device
11.26.63	3,112,035	S. G. Knight	Mobile Crane
03.02.65	3,171,545	S. G. Knight	Three Section Telescoping Crane Boom
08.09.66	3,264,950	S. G. Knight	Extensible Boom
08.09.66	3,265,220	S. G. Knight	Safety Control for Extensible Boom Cranes
08.16.66	3,266,638	C. Popov	Extensible Jib-Cranes
08.30.66	3,269,560	S. G. Knight	Safety Control for Extensible Boom Cranes
04.25.67	3,315,820	R. E. Stauffer	Crane with Winch Releasing Means
03.05.68	3,371,800	J. L. Grove	Safe Load Control Device for Cranes
07.29.69	3,458,053	G. Reuter	Cable Control Apparatus
02.23.71	3,565,402	Ernest A. Linke	Proximity Sensing Device
02.23.71	3,566,386	Martha Hamilton	Crane Angle Indicating System
08.24.71	3,601,259	John E. Olson	Crane Construction
04.25.72	3,658,188	G. P. Lamer	Safety Control for Load Line
08.01.72	3,680,714	T. M. Holmes	Safety Device for Mobile Cranes
03.20.73	3,721,350	O. T. Nephew	Boom Extension Control System
03.27.73	3,722,707	Nils E. Hedeen	Hydraulic Crane Control to Prevent Uncontrolled Falling of Load
04.03.73	3,724,679	Roy D. Brownell	Indicator or Control for Crane
04.03.73	3,725,887	V. R. Sneider	Radio Transmitting Alarm System
04.10.73	3,726,417	O. T. Nephew	Boom Extension Control System
06.05.73	3,737,888	Bernard J. Cheze	Method of Detecting an Overstepping of a Maximum Parameter Admissible for the Operation of a Machine
12.02.75	3,922,789	Ivan D. Sarrell	Boom Length Sensing System with Two-Block Condition Sensing
12.09.75	3,924,752	Frank M. Hoofnagle	Crane Boom and Line Safety Limit Warning Assembly
01.13.76	3,932,855	Martin W. Hamilton	Crane Radius Instrument
07.13.76	3,969,714	Gerald L. Greer	Safety System for Cranes
01.10.78	4,067,447	John B. Goss	Safety Device for Cranes

Patents for Limit Switches for Anti-Two-Blocking Devices

DATE	NUMBER	INVENTOR	ITEM
08.26.30	1,774,440	L. P. Hutt	Limit Switch
04.25.67	3,315,820	R. E. Stauffer	Crane with Winch Releasing Means

Suggested Design Criteria

Anti-two-blocking devices should be standard equipment, and retrofit programs should be developed to provide unequipped cranes with anti-two-blocking devices.

Since the first printing of this book in 1993, the occurrence of serious injury and death has become practically nonexistent from two-blocking because most cranes are now equipped with anti-two-block devices.

Representative Litigation

Over 120 two-blocking cases are known to have occurred that have resulted in death or very serious injury when the headache ball dropped on someone or the skip holding personnel was dropped.

Plaintiff Lyles; occurred 1975; Federal Court, Western District of Oklahoma, #C-76-0088-E.

> One of two boys walking 30 feet in front of latticework boom crawler crane was struck by falling choker and became a quadriplegic. Alleged the two-blocking lifted the weight of the boom from the pin-up lines and the vibration from travel caused the hoist line to become overloaded and break, causing the headache ball, hook, and choker in the hook to fall. Case settled during trial.

Plaintiff Lybek; occurred April 17, 1978; Circuit Court, Milwaukee County, Wisconsin, #467-265.

> Injured struck by falling bucket when headache ball allegedly two-blocked boom tip and parted the hoist line. Case settled.

Plaintiff Sidner; occurred July 15, 1981; Circuit Court, Escambia County, Florida, #82-50-CA-01.

> Flagman suffered brain damage when struck on head with headache ball when crane two-blocked. Verdict on behalf of plaintiff.

Plaintiffs Ferguson and Smith; occurred June 13, 1979; Federal Court, Oklahoma, #81-7932-D and 81-792-W.

> Two fatalities when lifting beam to clear building top during first shift after lunch with no flagman present. Plaintiff contended two hazards present: headache ball two-blocked, curled around boom tip and displaced straps from hook because of a Mickey Mouse latch. Defense verdict appealed and settled on behalf of injured.

Plaintiff Williams; occurred February 26, 1990; 234th Judicial District Court, Harris County, Texas, #90-54322.

> A fatality occurred when a worker was struck by a headache ball on a fly line (two hoist) mobile crane when extending the boom to position the main hoist line for lifting. Inoperative anti-two-blocking device in violation of manufacturer's instructions. Verdict on behalf of plaintiff.

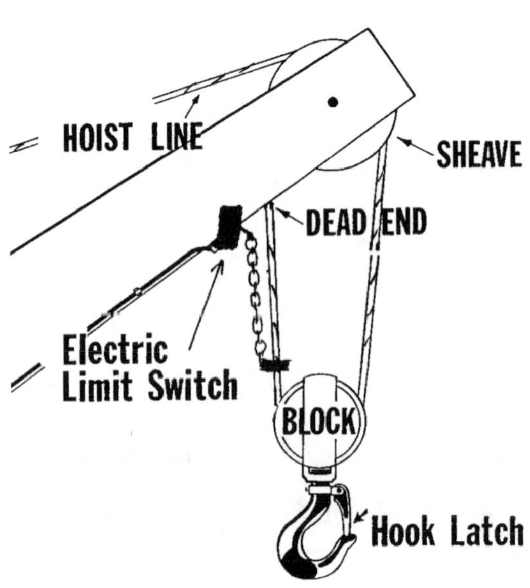

| **Figure 6-1** |

Anti-Two-Blocking Device

Operator's Station

No Crush-Resistant Cabs

Definition
When a crane with a light-weight, sheet-metal cab overturns, the operator may be crushed.

Description
Many crane cabs are not designed to withstand crushing in the event the crane overturns on the cab side, which often results in the operator being trapped inside and crushed. (See Figure 7-1.)

Risks Presented by Hazard
I have identified over 120 recorded occurrences in which operators were crushed in crane cabs. When a crane overturns on the left side where the operator's cab is generally located, and if the cab has been made of light-weight sheet-metal, the operator has no safe sanctuary and is usually crushed by the boom, the cab itself, or some other part of the crane. Because operators know most crane cabs are not crush-resistant, they often attempt to escape, which is a very unwise decision as they are usually killed or seriously crippled in the attempt.

Available Hazard Prevention Measures
In the 1950s it was recognized that protective canopies for heavy crawler-type bulldozers could be designed and fabricated that would resist the crushing effect of rollover. Beginning in the late 1960s, rollover protection system (ROPS) standards were developed by the Society of Automotive Engineers (SAE) for tractors (both crawler and wheel), loaders, graders, compactors, scrapers, water wagons, rear dumps, bottom dumps, fifth-wheel attachments, and other pieces of equipment. Deaths and crippling injuries from rollover have been substantially reduced because of ROPS and from falling objects because of FOPS. The same technology could be applied to cranes so operators would have the protection of a crush-resistant cab in the event of upset. Also see page 83 with the discussion on straddle cranes.

OSHA Requirements
OSHA 1926.1000 and 1926.1003 requires ROPS on tractors, front-end loaders, and scrapers but has not made this requirement applicable to crane cabs to assure for operator protection in the event of upset.

ANSI Requirements
ANSI has not commented on ROPS and crush-resistant cabs for cranes.

Other References
The Society of Automotive Engineers' *SAE Recommended Practices,* J1040c, "Performance Criteria for Rollover Protective Structures (ROPS) for Construction, Earthmoving, Forestry, and Mining Machines," sets forth

requirements for ROPS that could be adapted to cranes.

In 1970 the *Oregon Safety Code,* Chapter 16, "Logging Code," required rollover protection structures (ROPS) on crawler-type loaders. In the Northwest, track excavators are sometimes converted into log loaders, turning them into crawler-type loaders. These converted loaders need crush-resistant cabs to protect the life of the operator in the event of upset.

Suggested Design Criteria

The technology of rollover protection should be transferred to cranes to provide crush-resistant cab structures consistent with W2 design criteria — to support twice the weight of the crane at any point — developed by the U.S. Army Corps of Engineers and *SAE Recommended Practices,* J1040c. Operator restraint systems should also be provided. Considering the number of upsets, retrofitting cranes with crush-resistant cabs and restraint systems is a reasonable option to assure for a safe operator compartment.

Representative Litigation

Plaintiff Okeson; occurred June 24, 1983; U.S. District, Western District, Pennsylvania, #85-1606.

Deceased operating dragline while sluing loaded bucket. When catching bucket, dragline upset and he was trapped and crushed in cab of crane. Alleged that a crush-resistant cab was needed. Case settled.

No ROPS on Mobile Rough-Terrain Cranes

Definition

Mobile, rough-terrain cranes are particularly prone to upset in the travel mode because they have a high center of gravity. They should be equipped with crush-resistant cabs or rollover protective structures (ROPS).

Description

Mobile, rough-terrain cranes and crawler tractors can both be driven on rough ground. Rough-terrain cranes may upset on 35 degree slopes; crawler tractors are stable on slopes up to 57 degrees. ROPS are required on crawler tractors but not on the more unstable rough-terrain cranes. These cranes can also upset when making sharp turns or when on uneven

Figure 7-1 **Crushed Cab**

surfaces. On some makes of mobile hydraulic, rough-terrain cranes that have two-wheel/four-wheel steering, ROPS would provide protection for the operator in the event the operator inadvertently engages the four-wheel steering lever, cutting the turning radius in half and causing the crane to upset. (See Figure 7-2, Control Levers for Two-Wheel/Four-Wheel Steer.)

Risks Presented by Hazard

Death or serious injury can occur on mobile, rough-terrain cranes because of their propensity to overturn, crushing the lightweight, sheet-metal operator's cab. Even at moderate road speed of ten to fifteen miles per hour, well below their attainable gear and throttle settings, these cranes can upset when making a tight turn.

The great versatility provided by two-wheel/four-wheel steering increases the opportunities for upset as it is easy to drive a crane with this type of steering onto slopes, soft ground, or into deep ruts. If this type of top-heavy crane upsets to the left side where the cab is located, the operator has little chance of escape and is usually killed because the light-weight, sheet-metal cab gives no protection and is usually crushed between the boom and the ground.

Available Hazard Prevention Measures

1. Install after-market, custom-made crush-resistant cabs on cranes of this type.
2. A recent front-page article, "Legal Maneuvers," in the *Wall Street Journal* (January 5, 1993), reports that automobile manufacturers of the popular, sporty 4x4 Jeep-type vehicles with built-in roll bars rely upon a stability index that divides the width between the wheels by the height of the center of gravity. The achievable target rating of 2.25 is considered a reasonable measure of stability to reduce the risk of rollover during sharp turns at twenty to thirty miles per hour or more.

OSHA Requirements

OSHA has not commented on rollover protection for rough-terrain cranes and inadvertent movement of controls.

ANSI Requirements

ANSI has not commented on rollover protection for rough-terrain cranes and inadvertent movement of controls.

Other References

1. David V. MacCollum's chapter, "Rollover Protective Systems (ROPS)," *Automotive Engineering and Litigation*, Volume 2, Chapter 1, George A. and Barbara J. Peters, eds. (Garland Law Publishing, New York, 1988).
2. See Chapter 10, Controls, for references on inadvertent control activation.

Suggested Design Criteria

Crush resistant cabs need to be provided on all rough-terrain cranes. See SAE 1040(C), Performance Criteria for Rollover Protective Structures (ROPS) for Construction, Earthmoving, Forestry, and Mining Machines, as a basic starting point for development of a suitable cab protective structure.

Designers should assure that the four-wheel-steer control is well removed from other levers and in a place where inadvertent activation would be improbable.

Operator's manuals for rough-terrain cranes should state the stability index and the speeds at which sharp turns could cause upset.

| *Figure 7-2* |

Control Levers for Two-Wheel/Four-Wheel Steer

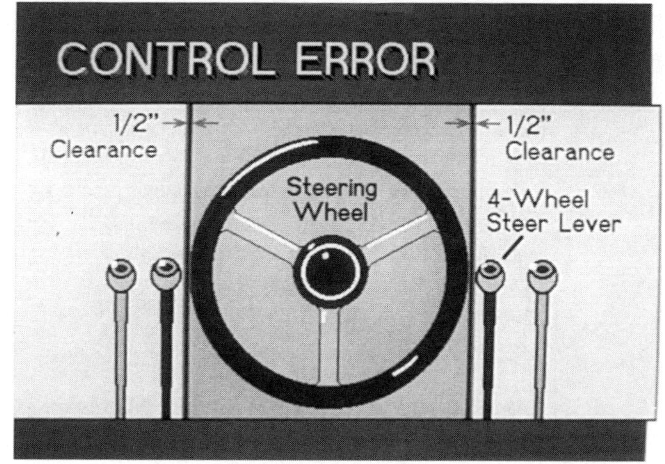

Representative Litigation

Plaintiff Skinner; Filed 1977; Circuit, Douglas County, Oregon, #77-0360.

> Deceased mechanic driving rig down slope to shop at night lost control, perhaps by striking four-wheel steer control lever. Went too close to bank, and crane rolled over. No pin in turntable. Alleged no crush-proof cab to protect the operator. Case settled.

Plaintiff Glerup; occurred January 11, 1985; State Court, Grays Harbour, Washington, #85-251-5.

> Deceased driving hydraulic crane on highway to sawmill. Went around a curve, off the road, and overturned, crushing operator. Alleged that a crush-proof cab was needed. Also alleged that inadvertent control activation may have been involved. Case settled.

No Operator Seat Belt

Definition

For the crane cab to be a safe refuge for the crane operator, a seat belt is necessary to prevent ejection and to deter the operator from making the unwise choice of attempting to escape from the cab.

Description

There are circumstances when crane operators are confronted with unstable cranes that start to tip and cause ejection from the cab that results in injury or death.

Risks Presented by Hazard

Crane operators are often confronted with a no-win situation when a crane starts to tip. They can remain in the cab and face possible ejection, or they can try to escape, which is usually a deadly choice. Operators are given no safe alternatives, but would stand a better chance of survival by remaining in the cab, especially if the cab were structurally enhanced. (See previous section: No Crush-Resistant Cabs.)

Available Hazard Prevention Measures

Provide a restraint system and handholds and warn that the best chance for survival in the event of upset is to stay within the cab (that hopefully has been made structurally sound), use the restraint system, and brace yourself with the handholds.

OSHA Requirements

OSHA has not commented on this issue.

ANSI Requirements

ANSI specifies the following:

> A seat belt shall be provided in all single control station wheel-mounted cranes for use during transit and travel. (Section 5-1.8-(e) Cabs, in Addenda of B30.5, December 30, 1992)

Other References

Seat belts for cranes was the topic of a Society of Automotive Engineers (SAE) standards committee meeting.

Suggested Design Criteria

Cranes should have restraint systems in their cabs, safety handholds, and warning labels with specific instructions on how to respond to crane upset. (See *Suggested Design Criteria* in next section: Emergency Exit Systems.)

Representative Litigation

Plaintiff Easton; occurred May 23, 1986; Civil District, Arlee's Parrish, Louisiana, #86-11586.

> The crane operator was crushed to death when he attempted to escape or was ejected from the crane cab when it upset. It was alleged that a warning and a restraint system was needed as the upset did not collapse the cab and that interlocks were needed to restrict boom movement when outriggers were retracted. Verdict for injured.

Plaintiff Anderson; occurred November 16, 1978; Eastern District, Northern Division, Missouri, #N79-0039-C.

> Operator lost left leg when attempting to escape as crane toppled when lifting a small reel of wire rope. Plaintiff contended that other earlier similar occurrences showed a need for a safe operator sanctuary. Plaintiff also alleged that interlock switches were needed to

limit boom movement when outriggers were retracted. Settled.

Plaintiff Noe; occurred June 29, 1982; Superior Court, Los Angeles County, California, #C-447-007.

Deceased crane operator attempted to escape overturning crane and fell 50 feet to his death. Alleged no instruction that he should stay with crane. Plaintiff settled with contractor. Defense verdict for manufacturer.

Plaintiff Spaniola; occurred June 16, 1987; 17th Judicial Circuit; Broward County, Florida, #88-18470-CO.

Injured sustained serious brain damage when thrown from the open cab of the truck-mounted crane being driven to work site with latticework boom assembled and extended over the rear of the truck carrier. The crane operator remained in the cab so that the boom could negotiate narrow streets. When traveling on the freeway, with doors open for ventilation in hot June weather, he fell from moving crane. Plaintiff is contending no seat belts, no communication system between truck driver and operator, and a solid window in door that could not be opened separately. Case settled.

Emergency Exit Systems

Definition
If a crane upsets, catches on fire, or the doorway is blocked by falling materials, the operator may be trapped in the cab.

Description
An upsetting crane can overturn into water. Other circumstances, such as fire or falling materials, may make it necessary for the operator to get out of the cab immediately. If the doorway cannot be opened, the operator is in trouble.

Risks Presented by Hazard
An operator can drown in a crane upset into water if the exit door happens to land face down in the water and no emergency exist system has been provided.

Available Hazard Prevention Measures
Select a crane that has an alternate exist system, such as quick release windows.

OSHA Requirements
OSHA has not commented on this subject.

ANSI Requirements
ANSI has not commented on this subject.

Other References
The Society of Automotive Engineers' (SAE) J185, "Access Systems for Off-road Machines," June 1988, describes alternate systems when a primary access is not usable.

Suggested Design Criteria
School buses, recreational vehicles, and aircraft have emergency escape systems that have long been recognized as an essential design consideration.

Emergency quick release systems for the front window, side windows, and rear windows of a crane are needed.

Cabs should be designed to resist distortion to prevent disabling of the quick-release escape mechanism.

An explicit and conspicuous label should be provided in a crane cab that gives instructions on how to operate the emergency escape system. These instructions should also be included in the operator's manual.

Figure 7-3
No Emergency Quick Release Exit

Metal grate on front of cab prevented escape after rollover, resulting in a fatality.

Representative Litigation

Plaintiff Halligan; Occurred March 10, 1981; District, Converse County, WY, #CA 8751.

Deceased drowned when crane overturned in pond. No emergency escape system for operator. Case settled.

Flying or Swinging Objects

Definition

Crane operators can be struck by flying or swinging objects that pierce the windshield or intrude into the cab.

Description

In demolition, pavement-breaking, scrap-iron handling, log-loading and other operations that require use of cranes, excavators, or other boomed equipment, operators are often imperiled by flying and swinging objects.

Risks Presented by Hazard

Because crane operators sit in fixed, confined locations, they do not have time to duck or avoid a flying or swinging object that, without warning, can pierce the windshield or intrude into the cab. The operator is, in essence, a sitting duck, completely vulnerable to being struck by any flying or swinging object.

Available Hazard Prevention Measures

When it is foreseeable that flying or swinging objects will be encountered, screens or grills over the windshield and windows should be installed.

Install safety glass or Plexiglas that is of bullet-proof thickness.

OSHA Requirements

OSHA has not commented on this hazard.

ANSI Requirements

Starting in ANSI Z26.1-1977 and Z26.1a-1980, Safety Glazing Materials for Glazing Motor Vehicles Operating on Land Highways, provides guidance.

Other References

SAE Recommended Practices, starting in J673, "Automotive Safety Glazing" (November, 1983), and J674, "Safety Glazing Materials-Motor Vehicles" (November, 1990). The B30.5 Addenda (March 30, 1990), in Section 5-1.8(e) Cabs states.

All cab glazing shall be safety glazing material as defined in ANSI Z26.1. Windows shall be provided in the front and on both sides of the cab or operator's compartment with visibility forward and to either side. Visibility forward shall include a vertical range adequate to cover the boom point at all times. The front window may have a section which can be readily removed or held open, if desired. If the section is of the type held in the open position, it shall be secured to prevent inadvertent closure. A windshield wiper should be provided on the front window.

Suggested Design Criteria

Accessory cab protection should be designed and provided to protect operators from flying and swinging objects. Such cab protection is particularly necessary for cranes used consistently in demolition, pavement-breaking, scrap-iron handling, log-loading, etc.

Representative Litigation

Plaintiff Mozingo; Filed 1989; U.S. District, Southern District of Ohio, Eastern Division, #C2-89-432.

Excavator operator was blinded in one eye when using a hydraulic rock breaker attached to the crane boom, causing a rock fragment to fly through the tempered glass cab window. Alleged need for safety glass and protective grill for such activities. Case settled.

Access

Unsafe Access to Cab, Housing and Cab Roof

Definition

When inadequate steps, ladders, handholds, railings, and dangerous walkways are present, the hazard of unsafe access exists.

Description

The design of many cranes has not taken into consideration the need for crane operators, oilers, maintenance personnel, and others to have safe entry and exit into and from their particular work areas on the crane.

Risks Presented by Hazard

Often access onto and off cranes is inconvenient, difficult, and unsafe. When those who must work in and about a crane are provided with only marginally safe access due to inadequate ladders, steps, walkways, and platforms, the risk of falls from an elevation or on the same level is great.

Access is also hazardous if the smooth surfaces over which people must walk have become coated with oil, grease, ice, and mud, making access slippery.

Access is also made unsafe if ladders, steps, walkways, and platforms do not have sufficient handholds and railings.

Access into certain crane locations can be dangerous if hot exhaust pipes or other obstacles are located in and about the normal entry and exit pathways.

Available Hazard Prevention Measures

Non-slip steel treads, surfacing materials, and a wide variety of handholds are commercially available, leaving little reason for having unsafe access on and off cranes.

Exhaust pipes should be guarded or insulated to prevent contact by people who are gaining access to or working on or about the crane. Such exhaust systems should be located to avoid hazardous, toxic fumes from accumulating in the operator's cab.

OSHA Requirements

1. OSHA 1926.550(a)(10) and (11) assures for guarding of exhaust pipes and dispersion of exhaust fumes from the operator's cab.
2. OSHA 1926.550(a)(13) assures that guardrails, handholds, and steps should conform to ANSI B30.5, *Crawler, Locomotive and Truck Cranes*.
3. OSHA 1910.179(c)(1) and (2), Marine Terminal Standard 1917.45(f)(4), and OSHA 1926.550(a)(13) covers location and other access requirements for crane cabs, bridges, and/or runways.
4. OSHA 1910.132(a) covers safety requirements for safety belts, lifelines, and lanyards.

ANSI Requirements

ANSI's A14.3, *Safety Requirements for Fixed Ladders*, addresses access requirements. Individual standards for each type of crane also address access.

Other References

The Society of Automotive Engineers' *SAE Recommended Practices*, J185, "Access Systems for Off-Road Machines" (first approved in 1970), gives excellent design criteria for access systems for equipment such as cranes.

Suggested Design Criteria

Crane design should incorporate the Society of Automotive Engineer's *SAE Recommended Practices,* J185, "Access Systems for Off-Road Machines." Vertical access handholds, steps or ladder rungs should easily accommodate individuals climbing up or down and assure that contact of either two-feet-and-one-hand or two-hands-and-one-foot can be made on steep steps and ladders at all times.

Exposure of those climbing on or off a crane to hot exhaust pipes should also be considered.

Materials used for accessways should be of the expanded-metal or other type that will prevent accumulation of fuel, oil, or grease from the crane.

Representative Litigation

Many cases have involved slips and falls on or off cranes. Juries find for the injured when access is so unsafe that only the agile can gain access, especially when informed of the very explicit standards for safe access that exist and were not applied. (See Figure 8-1.)

Plaintiff Freeman; occurred July 29, 1975; Judicial Court, Duval County, Florida, #77-8735-CA.

> Injured fell when attempting to get off crane. Access was oily. Alleged there was no clear route of access consistent with SAE standards. Case Settled.

Plaintiff Phillips; Filed 1982; 3rd Judicial District Court, Madison County. Iowa, #81-L-1113.

> Injured's foot mangled when boarding crane and had to step on wheel which began moving backwards. Alleged unsafe access. Case settled.

Figure 8-1 **Unsafe Access**

Unsafe Walkways on Overhead Bridge Cranes

Definition
Track rails or other walkways often fail to provide safe access to overhead cranes.

Description
How an operator must gain access to the crane cab is often ignored in overhead crane design. To get to the cab, operators sometimes have to walk on either of the single rails on which the bridge crane runs and that sit on opposite sides of a building. Some of these cranes have a walkway on the bridge to provide access to the operator's cab. Maintenance and service people may also have to use these means of access. Lighting of access routes is sometimes very poor.

Lights to illuminate the lifting area are sometimes suspended from trap doors on the floor of the walkway running along one side of the bridge to provide access to the operator's cab and for maintenance of the trolley hoist machinery. These trap doors must be opened and raised in order to replace burned-out lights.

Risks Presented by Hazard
Operators of overhead cranes are very vulnerable to falls when access on and off the crane is poorly lighted and no special walkways with handrails are provided.

In one instance, the trap door on which lights were suspended was unintentionally left open and unattended when a light bulb needed replacing. Another person on the night shift who was unaware of the unattended opening came along and fell through the trap door to his death sixty feet below. The walkway was not illuminated and the open trap door was not visible.

Maintenance and service is often done by outside contractors or suppliers. If one of their personnel is injured due to unsafe access, the owner of the facility may be found liable.

Available Hazard Prevention Measures
Provide safe walkways with handrails for access to the crane cab and maintenance areas on overhead bridge cranes. (See previous section: Unsafe Access to Cab, Housing and Cab Roof.)

To avoid falls through open trap doors, change design to eliminate lights suspended from the bottom side of such trap doors.

Provide lighting of an intensity of at least ten foot-candles.

OSHA Requirements
OSHA 1910.179(c)(2) and (4) define requirements for safe access to cranes and lighting.

ANSI Requirements
1. ANSI A14.3, *Safety Requirements for Fixed Ladders*, clearly illustrates additional methods to achieve safe access.
2. ANSI A12.1, *Safety Requirements for Floor and Wall Openings, Railings, and Toeboards*, provides additional guidance for overhead and gantry cranes.
3. ANSI B30.2.0, *Overhead and Gantry Cranes*, Section 2-1.5.3, states: "… no step over any gap exceeding 12 inches (304.8 mm)."

Other References
1. *Safety and Health Requirements Manual*, EM 385-1-1 (U.S. Army Corps of Engineers, November 3, 2003). Table 7-1 requires an intensity of ten foot-candles for walkways, ladders, and stairs.
2. The National Safety Council provides guidance on how to achieve safe access in its *Accident Prevention Manual for Industrial Operations, Engineering and Technology*, Ninth Edition (National Safety Council, 1988).

Suggested Design Criteria
Designers who oversee the installation of custom-built overhead bridge cranes need to assure that access walkways for both the operator and maintenance personnel are consistent with ANSI requirements for both fixed ladders and walkways and are properly lighted.

Representative Litigation
Plaintiff Carlini; occurred March 6, 1987; Court of Common Pleas, Allegheny County, Pennsylvania, #88-9358.

Crane operator, when taking a break, fell approximately sixty feet through an open trap door. The purpose of the trap door was to be able to service a light suspended from its bottom side. There was no other overhead lighting in the foundry, making the walkway very dark. Maintenance people had opened the trap door and left it open and unattended. Alleged that placement of lights on bottom of trap doors was unsafe. Case settled.

9

Pinch Points and Nip Points

Pinch Points

Definition

Any point at which it is possible to be caught between the moving parts of a machine, between moving and stationary parts of a machine, or between the material being worked and the moving parts of a machine (*Dictionary of Terms Used in the Safety Profession*, Fourth Edition, American Society of Safety Engineers, 2001).

Description

On cranes, two giant pinch points exist:

1. The accessible area within the swinging radius of the rotating superstructure of a crane (the cab/counterweight) that people may inadvertently enter in the narrow clearance between the rotating cab/counterweight and its stationary carrier frame. (See Figure 9-1.) Analysis of occurrences shows that victims were usually "invited" into the danger zone for reasons such as these:

   a. access to the water jug

   b. access to the tool box

   c. access to the outrigger controls

   d. access to perform maintenance

   e. access to storage of rigging materials

   f. access to the flat surface on the carrier bed often used by supervisors, foremen and others as a place to lay out drawings for momentary jobsite review

2. When a crane is used in a confined space with close clearance, the accessible area between the cab/counterweight and any stationary object such as a building, wall, or parked vehicle. Invariably injury occurs when a crane is located in an accessway that invites people to walk by it to reach another work area.

This hazard is inherent in many mobile cranes, especially rough-terrain, truck-mounted, and crawler cranes. Many people, especially oilers, have been inadvertently crushed when caught in such pinch points.

In all of the known cases where someone entered the danger zone and was caught in a pinch point, the danger zone was outside the crane operator's vision. Survivors of these occurrences have consistently stated they had good cause to believe the crane operator was not going to rotate or slue the boom at that particular moment.

Risks Presented by Hazard

Since before World War I, the serious consequences of this hazard have been known. The crushing or squeezing inherent with this hazard has caused serious crippling injury and even loss of life.

Available Hazard Prevention Measures

Some models of cranes have been designed to meet the first order of precedence in system safety to remove or eliminate the hazard by sloping the bottom of the cab from where it meets the turntable upward at approximately 30 degrees and leveling it out at a point 7½ feet above the ground where the counterweight is normally located. The likelihood of anyone's being caught between the sloping underside of the cab and the carrier frame is negligible, as they would be pushed aside rather than crushed. The pinch point created between the cab and a stationary object would also be eliminated, as the cab would be well above the head level of anyone passing by to reach another work area. Even a clearance of 14 to 16 inches between the cab/counterweight and the carrier frame or crawler tracks would render the pinch point between the cab and carrier frame relatively harmless. The selection of a crane that provides safe clearance and eliminates these pinch points reduces workplace hazards.

Remove water jugs, tool boxes, and rigging materials from this dangerous area to reduce the incentive to enter the danger zone.

Install rear view mirrors for the crane operator to provide a redundant safeguard so the operator can see into the turning area of the cab/counterweight.

Erect barriers or fences around crane operations to prohibit entry into the danger zone.

Install warning labels in the pinch point hazard area to (1) alert people entering that area to get out of the danger zone, and (2) illustrate the appropriate methods of barricading the pinch point danger zone. (See Figure 9-2.)

OSHA Requirements

1. Subpart N, 1926.550(a)(9), requires: "Accessible areas within the swing radius of the rear of the rotating super-structure of the crane, either permanently or temporarily mounted, shall be barricaded in such a manner as to prevent an employee from being struck by the crane."
2. OSHA 1910.180(i)(6) requires that locomotive cranes should not be swung when the possibility exists that they might strike other railroad cars on adjacent tracks and defines the need to barricade the swing area effectively to avoid striking people with the crane cab and counterweight.
3. Marine Terminal Standard 1917.45(i)(2) and Longshoring Standard 1918.74(a)(10) and 1918.74(c) address guarding of the swing radius.

ANSI Requirements

1. ANSI B30.15, *Mobile Hydraulic Cranes*, does not address this subject.
2. ANSI B30.5, *Mobile and Locomotive Cranes*, Section 5-3.4.4. addresses the hazard of swinging into adjacent railroad cars.

Other References

1. *Safety and Health Requirements Manual*, EM 385-1-1, U.S. Army Corps of Engineers, published November 3, 2003, Section 16.C.09 requires the following clearances:

   (1) Adequate clearance shall be maintained between moving and rotating structures of the crane and fixed objects to allow the passage of employees without harm: the minimum adequate clearance is 16 inches.

   (2) Accessible areas within the swing radius of the rear of the rotating super-structure of a crane, either permanently or temporarily mounted, shall be barricaded to prevent an employee from being struck or crushed by the crane.

2. *Crane Handbook* by D. E. Dickie of the Canadian Construction Safety Association of Ontario (1975) stresses the importance of barricading the swing and working radius of the crane on p. 47 and in Figures 2.13, 3.2, and 3.5.

3. *Accident Prevention Manual for Industrial Operations* (National Safety Council), on p. 557 of the 6th Edition and on p. 688 of the 7th Edition, has an illustration of crane guarding that prevents a person from being caught in the shear point between the crane cab and carrier body.

4. *Industrial Accident Prevention* by David Stewart Beyer, Ph.B. (Houghton

Mifflin, 1916) states on p. 182 in a section entitled "Locomotive Cranes":

> "Truck Frame. A clearance of at least 18 inches shall be provided between the bottom of the boiler and the truck frame."

5. Model Code of Safety Regulations for Industrial Establishments for the Guidance of Governments and Industry published by the International Labour Office, Geneva, Switzerland, in 1949. Page 275, Regulation 156, Locomotive Cranes, paragraph 4, "Body Clearance," reads:

> "4. A clearance of at least 35 cm (14 in.) shall be provided between rotatable bodies of locomotive cranes and the frames of the crane trucks, in order to prevent workers from being crushed against the frames if caught by the revolving body."

6. The following U.S. patents indicate that people have been aware of this hazard and have designed preventive devices:

DATE	NUMBER	INVENTOR	ITEM
12.30.24	1,521,057	Charles Trezona	Safety Apron Construction for Rotating Platforms
10.20.25	1,557,632	John S. Tresider	Safety Device for Revolving and Clarence Boss Platforms
08.30.27	1,640,677	William E. Smiley	Safety Appliance
09.11.34	1,973,445	Robert S. Quinn	Guard Device for Moving Structures
03.24.36	2,035,044	G. E. Carle	Safety Device for Cranes
08.08.44	2,355,361	J. H. Brown	Safety Device
06.30.59	2,892,549	Budd W. Andrus	Safety Guards for Cranes and the Like (See Figure 9-3.)
05.28.74	3,812,978	David H. Roland and Edward Hastings	Crane Safety Barrier (See Figure 9-4.)
03.06.84	4,434,901	Paul O. Gehl	Safety Apparatus for Cranes

Suggested Design Criteria

Future crane design could slope the bottom of the cab from where it meets the turntable upward at approximately 30 degrees and level it out at a point 7½ feet above the ground where the counterweight is normally located. Even a clearance of 14 to 16 inches between the cab/counterweight and the carrier frame or crawler tracks would reduce the hazard.

New crane design should include a built-in guard system on the crane cab and counterweight to bar entry to the rotation area. Retrofit kits with similar guards should be provided to fit existing cranes.

Elimination of storage areas for tools, water, and other items from the danger zone so there is no need for anyone to go into this area.

Available technology can remove this hazard by using these two devices:

- Closed circuit TV monitors so the operator is able to view the danger zone.

- Infrared detectors that sound an alarm if a *warm* body is present in the danger zone to let the operator know about someone's presence before he makes a swing so he can visually check in the TV monitor to see that personnel are in the clear.

Many cranes have warning labels to warn of this danger. However, if closed circuit TV monitor and infrared detectors are not available, the area to the rear of the crane should be barricaded.

Representative Litigation

Plaintiff Emmons; occurred July 10, 1981; Circuit Court, Mobile County, Alabama, #CV-82-001548.

> Counterweight of crane caught injured behind head and pulled him off the ground and up, underneath and between the counterweight and the main part of the truck carrier. No rear-view mirror, no warning, and narrow clearance between cab and counterweight. Employer placed

water container where it would lure workers into danger zone created by swinging cab. Plaintiff alleged need for greater clearance between truck carrier and crane cab. Case settled.

Plaintiff Sadler; occurred January 25, 1980; U.S. District, Eastern District, Missouri, U.S. District, Eastern District, Missouri, #S82-0087C.

Oiler crushed while cleaning tracks to avoid frozen mud when operator slued crane boom around, causing oiler to be struck by the counterweight and caught and crushed between the track and counterweight. Plaintiff alleged need for greater clearance between dragline cab/counterweight and crawler tracks. Case settled.

Figure 9-1 **Close Clearance Between Crane Cab and Truck Carrier**

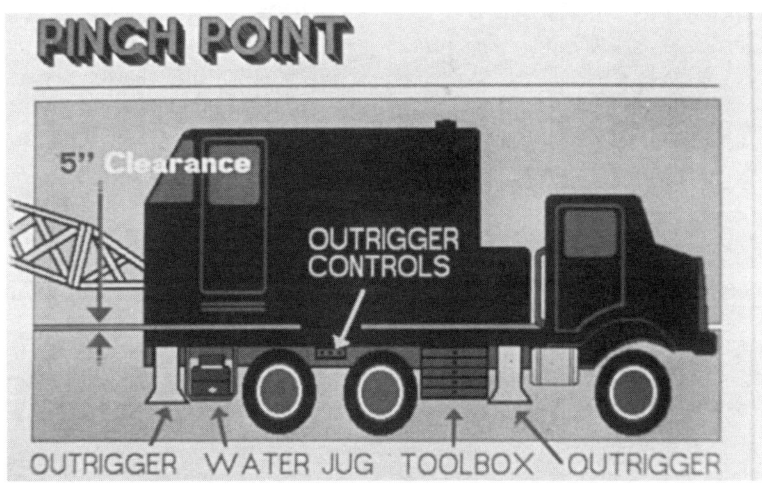

Figure 9-2 **Warning Label for Shear Point**

| *Figure 9-3* **U.S. Patent 2,892,549** |

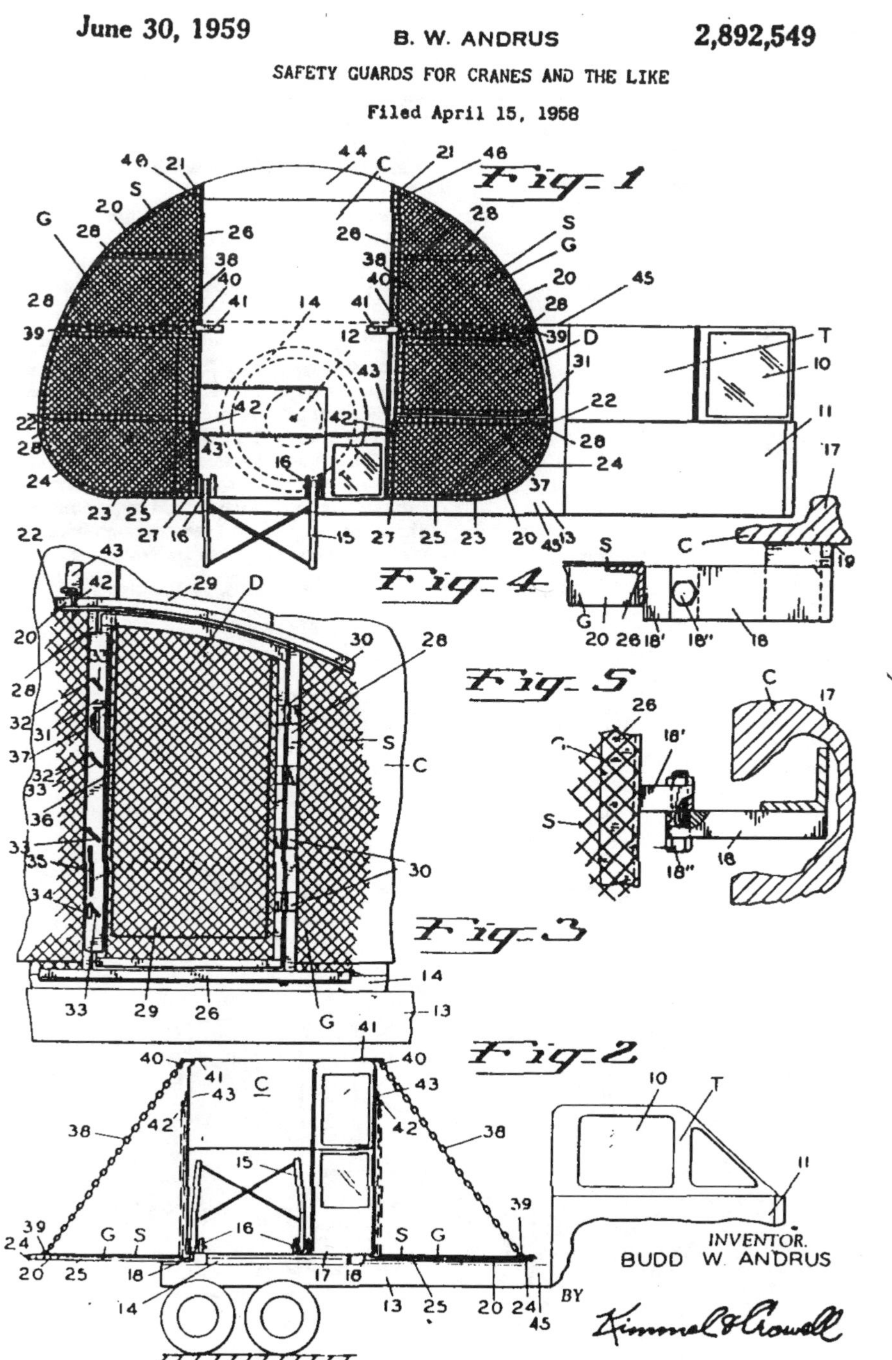

Close Clearance Between Crane Cab and Truck Carrier.

| *Figure 9-4* **U.S. Patent 3,812,978** |

United States Patent [19]

Roland et al.

[11] **3,812,978**

[45] **May 28, 1974**

[54] **CRANE SAFETY BARRIER**

[75] Inventors: **David H. Roland; Edward D. Hastings, both of Charlotte, N.C.**

[73] Assignee: **J. A. Jones Construction Company, Charlotte, N.C.**

[22] Filed: **Nov. 21, 1972**

[21] Appl. No.: **308,545**

[52] U.S. Cl. 212/1, 256/1
[51] Int. Cl. .... B66c , E04h 17/00
[58] Field of Search .... 212/1, 28; 280/150 R; 256/10, 1; 52/73; 74/60R, 609, 111; 272/39, 42

[56] **References Cited**
UNITED STATES PATENTS

51R,642	4/1894	Breen	280/150 R
2,704,498	3/1955	Furnas	280/150 R
2,856,159	10/1958	Braddock	256/10

Primary Examiner—Evon C. Blunk
Assistant Examiner—H. S. Lane
Attorney, Agent, or Firm—Clarence A. O'Brien; Harvey B. Jacobson

[57] **ABSTRACT**

A barrier supported from the base of a crane for movement therewith and shiftable between retracted and horizontally outwardly extended position relative to the base of the crane for enclosing an area about the crane base within which the oscillatable superstructure of the crane is swingable, so as to thereby exclude accidental movement of persons into the area about the base of the crane through which portions of the crane superstructure are swingable.

10 Claims, 14 Drawing Figures

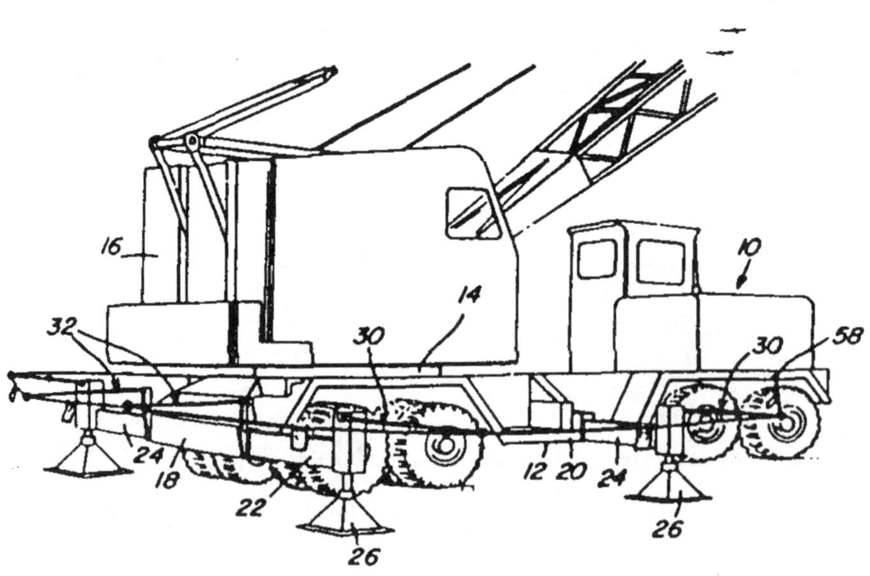

Crane Safety Barrier.

| *Figure 9-5* **Design Concept to Prevent Pinch Point Injuries** |

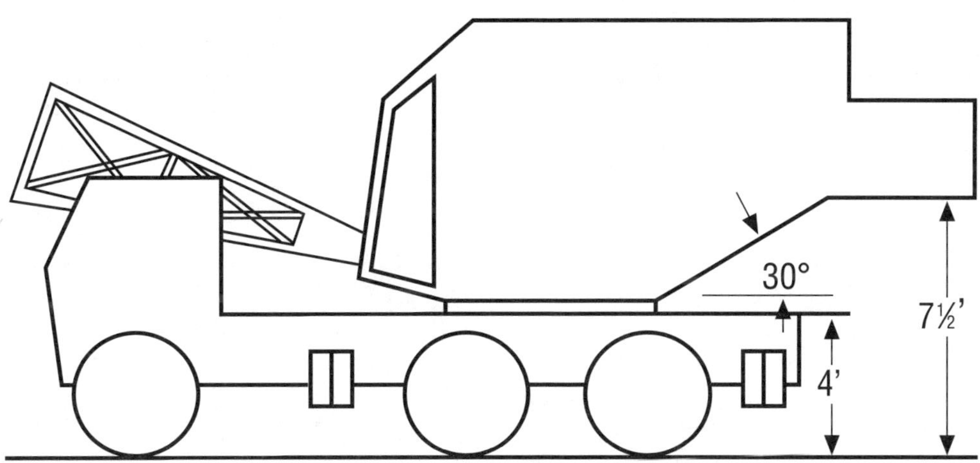

Design Concept to Prevent Crane-Cab/Carrier-Frame Pinch Point Injuries.

Unguarded Nip Points

Definition

In *Dictionary of Terms Used in the Safety Profession,* Fourth Edition (American Society of Safety Engineers, 2001) a nip point is defined, in machinery and mechanical equipment, as the point of intersection or contact of two opposed circular surfaces, or a plane surface and a circular surface in which a worker's body or clothing may be caught.

Description

This hazard includes all unguarded moving or rotating parts within the crane cab, engine compartment, or service area that are accessible to people who must enter or reach into the crane's mechanical housing or compartment.

Risks Presented by Hazard

It has been assumed by some designers that access doors and panel covers serve as guards to keep people away from moving parts and that cranes are always shut down when adjustments or servicing is to be done. These assumptions are almost always wrong. Because it is so handy, the cab is commonly used by maintenance and other people as a convenient storage space for oil, spare parts, tools, lunch boxes, clothing, and other items. Oilers, maintenance personnel, and crane operators are often injured within the crane cab when caught by one of the many unguarded moving parts. The cramped machinery space inherent in crane cabs makes it easy to become entangled with unguarded moving parts. Loss of fingers, hands, or feet is very common when moving parts are unguarded. Such parts wait for the unwary to become entrapped.

Available Hazard Prevention Measures

All moving parts that are accessible to people need their own guards, even though inside the cab or a compartment.

OSHA Requirements

1. OSHA 1910.179(e)(6)(i), (ii), and (iii) for overhead and gantry cranes states: "Exposed moving parts such as gears, set screws, projecting keys, chains, chain sprockets, and reciprocating components which might constitute a hazard under normal operating conditions shall be guarded...."

2. OSHA 1926.550(a)(8) states: "Belts, gears, shafts, pulleys, sprockets,

spindles, drums, fly wheels, chains, or other reciprocating, rotating, or other moving parts or equipment shall be guarded if such parts are exposed to contact by employees, or otherwise create a hazard…."

3. Marine Terminal Standard 1917. 45(e)(1) states: "When exposed moving parts such as gears, chains and chain sprockets present a hazard to employees crane and derrick operations, those parts shall be securely guarded."

ANSI Requirements

In 1996 additional explanatory information was developed. ANSI B15.1-1992 revised, *Safety Standard for Mechanical Power Transmission Apparatus,* covers guarding, and the December 1992 Addenda for B30.5 states:

5-1.9.6 Guards for Moving Parts

(a) Exposed moving parts such as gears, set screws, projecting keys, chains, chain sprockets, and reciprocating or rotating parts which may constitute a hazard under normal operating condition shall be guarded.

(b) Guards shall be fastened and shall be capable of supporting, without permanent distortion, the weight of a 200 lb (90 kg) person unless the guard is located where it is impossible for a person to step on it.

Other References

Concepts and Techniques of Machine Guarding, published by the U.S. Department of Labor in 1980, OSHA No. 3067, gives excellent insight on guarding.

Suggested Design Criteria

Cranes could be designed to comply with the intent of the current issue of ANSI B15.1, *Safety Standard for Mechanical Power Transmission Apparatus,* and U.S. Department of Labor's *Concepts and Techniques of Machine Guarding.* ANSI B15.8 is written to address stationary equipment. These same concepts should apply to mobile equipment such as cranes.

Representative Litigation

Plaintiff Hoskie; occurred approximately 1983; District Court, Albuquerque, New Mexico, #83-0561-B.

While inside of machinery house, female plaintiff became entangled in left side of unguarded gearing, losing both hands, while crane was in operation. Plaintiff alleged that moving parts should be guarded. Verdict for plaintiff.

Plaintiff Watkins; Filed 1986; 333rd Judicial District, Harris County, Texas, #86-51.

Injured was attempting to grease exposed unguarded gears in crane cab using an aerosol grease can. Injured's arm was caught and crushed by moving gears. Alleged moving parts should be guarded. Case settled.

10

Controls

Control Confusion/Inadvertent Activation of Controls

Definition

Non-uniform placement of controls and controls that can be inadvertently activated cause operator error. (See Figure 10-1.)

Description

Cranes, like automobiles, come in many different makes and models and are operated by many different people. But there is a difference. Brake and accelerator pedals and other basic controls on automobiles are always in the same place. People who rent different cars while traveling on business do not have a problem. On cranes, the location of controls often varies, so crane operators who often change from crane to crane in normal work circumstances are continually faced with having to operate controls located in different places. Cranes also do many more things than automobiles, requiring the operator to use many more controls concurrently than when driving a car. Crane operation is much more complex.

Inadvertent activation can arise when controls do not have a latching mechanism requiring two motions of the hand for activation. For example, a control lever needs to have a button to release the lever from the detent before the lever can be moved. Some controls can be activated by bumping, either inadvertently by the operator or by other means. See Chapter 7, which discusses inadvertent activation of four-wheel steer.

Inadvertent activation also occurs with levers that fail to return to a neutral locked position when the control lever is released by the operator.

Another hazard can be the absence of a necessary control. For example, a crane without a mechanized dog, locking pin or swing (house) lock control, see Figure 10-1, can result in unintended cab and boom rotation.

Risks Presented by Hazard

A multitude of serious circumstances can arise from unintentional use or activation of a control that can lead to upset, two-blocking, loss of load, etc., which in turn can lead to injury, death, or property damage.

Available Hazard Prevention Measures

Operator training is often the only option available to overcome the inherent hazards created by the configuration of a crane's controls. Even with the best and most well-trained operators there is still room for error.

Training should focus on the control irregularities and deficiencies of the crane to be used. Optimum training would include sitting in the crane and going over each control, making the operator aware of specific mistakes that can be made.

When possible, for optimum performance, crane operators should be assigned to the same crane or the same make and model to avoid having to work with dissimilar controls

A "what if" review of all controls and possible crane component movements should be made to assess if the controls function as intended and the movement is in the correct direction. It should then be determined if the crane will make any unintended motion by itself. If any of the above occur, the manufacturer should be promptly notified in writing! Warning!—many cranes now rely upon an automatic shift like an automobile's "hill-hold" to hold the load in power lowering. Whenever it slips, either the manufacturer or his authorized dealer should maintain it because adjustment of these devices is tricky. When maneuvering a load in close tolerances, an unexpected drop of several inches can be extremely hazardous.

OSHA Requirements

OSHA has not commented on this specific hazard. However, OSHA Subpart C, "General Safety and Health Provisions," 1926.20(a) and (b) sets forth requirements for a safe workplace. Marine Terminal Standard 1917.45(f)(1) discusses operating controls.

ANSI Requirements

ANSI standards for cranes have suggested basic control arrangements for new cranes.

Other References

1. The Society of Automotive Engineers' *SAE Recommended Practices*, J983, "Crane and Cable Excavator Basic Operating Control Operations" (October 1998), covers uniformity.

2. Additional requirements for avoiding inadvertent activation were started early on in:
   a. "Controllers (NI)," *Safety Code for Cranes, Derricks, and Hoists*, ASA B30.2-1943 (The American Society of Mechanical Engineers, 1952), p. 33, Paragraph 1043C.
   b. *Factories, The Construction (Lifting Operations) Regulations*, Statutory Instruments No. 1581 (Minister of Labour of England, 1961), p. 8.
   c. "Controls," *Human Engineering Guide for Equipment Designers*, Second Edition (1964), pp. 2-91.
   d. "Pedals," *Military Standard, Human Engineering Design Criteria for Aerospace Systems and Equipment*, MIL-STD-803A-1(USAF), (Department of Defense, January 27, 1964), p. 61.
   e. *Safety Code for Crawler, Locomotive and Truck Cranes*, USAS B30.5-1968, (American National Standard, 1968), pp. 9 and 10.
   f. *Military Standard, Human Engineering Design Criteria for Military Systems, Equipment and Facilities*, MIL-STD-1472 (Department of Defense, February 9, 1968), pp. 54 and 67.
   g. Conover, D. W. and W. E. Woodson, "The Design Implications of Product User Behavior" (The National Commission on Product Safety, Letter Contract No. 70-168, December 30, 1969), p. 125.
   h. "Crane-Shovel Basic Operating Control Arrangements," SAE J983, *1971 SAE Handbook, SAE Recommended Practice*, p. 1187.
   i. Chapanis, Alphonse, "Design of Controls," *Human Engineering Guide to Equipment Design* (1972), Chapter 8.
   j. *Military Standard, Human Engineering Design Criteria for Military Systems, Equipment, and Facilities*, MIL-STD-1472B (Department of Defense, December 31, 1974), pp. 64, 86 and 87.
   k. "Controls," *Military Standardization Handbook, Human Factors Engineering Design for Army Materiel*, MIL-HDBK-759 (March 12, 1975), pp. 51 and 52.
   l. Woodson, Wesley E., *Human Factors Design Handbook* (1981), pp. 571-643.
   m. Huchingson, R. Dale, "Design of Aerospace Controls and Control Systems," *New Horizons for Human Factors in Design* (1981), Chapter 5, pp. 145-187.

Suggested Design Criteria

Controls need to be designed so they require a two-hand motion for activation.

Controls need to be designed so that when a lever is released, it automatically returns to neutral.

When rebuilding a crane, the control system should be redesigned and rebuilt to conform to the Society of Automotive Engineers' *SAE Recommended Practices*, J983, "Crane and Cable Excavator Basic Operating Control Operations" and applicable ANSI standards to provide uniformity in control systems. This type of action would be a part of any program for continuous improvement of equipment.

Representative Litigation

Plaintiff Jones; occurred August 29, 1979; Alameda County, California, #H-70551-B.

Deceased, while guiding a load suspended on the whip line of the boom jib, was struck by the headache ball and hook on main hoist line when the operator inadvertently touched the main hoist lever, causing the headache ball and hook to two-block. Alleged failure to provide a control lever that required two distinct motions for activation and no anti-two-blocking device. Case Settled.

Plaintiff Arvizu; occurred December 28, 1980; Pima County, Arizona.

Deceased struck by block and hook assembly that over-reeled (two-blocked) on pendant-controlled bridge crane. Limit switches on bridge crane had been disconnected. Alleged that operator may have pushed wrong button. Case dismissed due to negligence of employer who was immune from liability under workmen's compensation.

| *Figure 10-1* **Typical Mobile Hydraulic Crane Cab Control Station** |

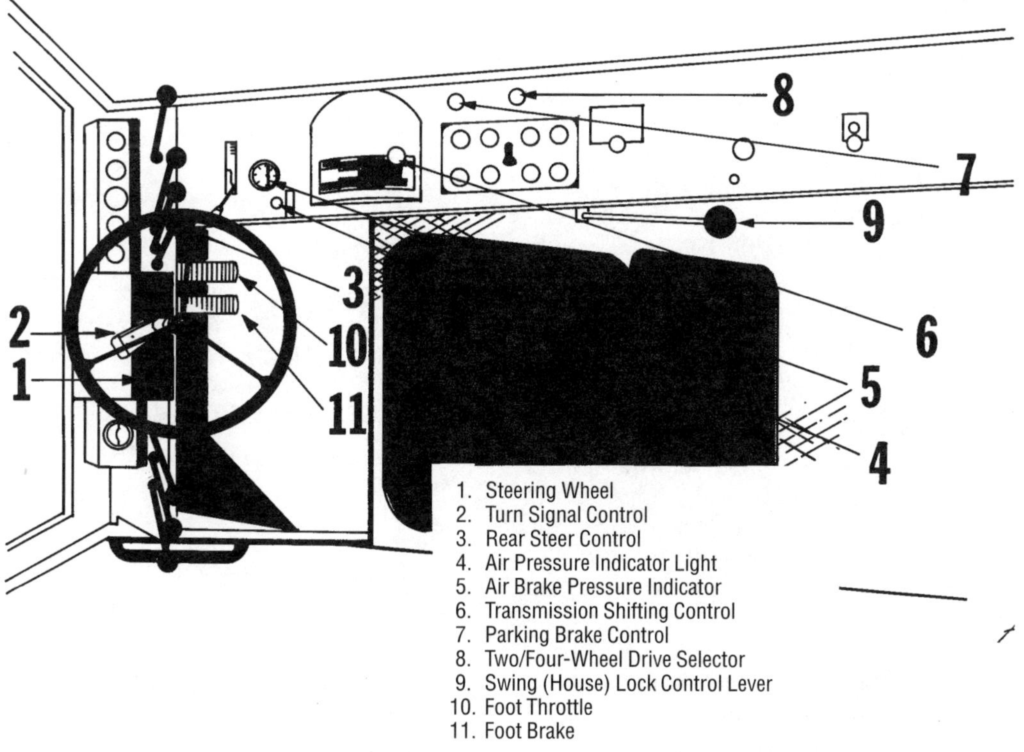

1. Steering Wheel
2. Turn Signal Control
3. Rear Steer Control
4. Air Pressure Indicator Light
5. Air Brake Pressure Indicator
6. Transmission Shifting Control
7. Parking Brake Control
8. Two/Four-Wheel Drive Selector
9. Swing (House) Lock Control Lever
10. Foot Throttle
11. Foot Brake

Plaintiffs Thompson, Terrell, Yodice, and Gandolfo; occurred March 21, 1981; Supreme Court, New York County, New York 07278/83.

Two killed, two seriously injured when joy stick control energized clamshell's closing motor, causing side pull on the clamshell bucket parked on mezzanine floor while the main hoist line was being used independently to raise spare parts to the mezzanine floor. No lockout switches available in crane cab to disconnect power to closing motor while maintenance tasks were done. Verdict for plaintiffs. Upheld on appeal.

Controls on a Conductive Tether or Controls Reachable from the Ground

Definition

Small flatbed-truck-mounted pedestal cranes often have hand-held remote controls on wire tethers or controls that can be reached when standing on the ground next to the truck. In either case, if the boom contacts a powerline, the operator can be electrocuted.

Description

Small flat-bed-truck-mounted cranes are used to load, transport, and off-load lumber, brick, and other building materials. Hand-held remote controls often have small toggle switches close together that can be inadvertently activated. (See Figures 10-2 and 10-3.)

Risks Presented by Hazard

Numerous deaths and crippling injuries from electrocution have been sustained by operators of these small cranes. Because this type of crane is often used at residential building sites to unload materials where convenient unloading may be under or next to powerlines, error-provocative controls on a conductive tether or controls that can be used while standing on the ground are especially hazardous. (See Figure 10-4.)

Available Hazard Prevention Measures

Select a radio remote control or other nonconductive control system.

Select a model that requires the operator to stand on the flatbed of the truck to operate the pedestal crane.

The use of a proximity alarm wired to prevent boom operation in the vicinity of powerlines would reduce the opportunity for the truck to be loaded or unloaded under or next to powerlines.

The electric utility should be asked to de-energize powerlines in the area where materials are being handled with boomed equipment. (See Chapter 4, Powerline Contact.)

OSHA Requirements

See OSHA Requirements in Chapter 4, Powerline Contact.

ANSI Requirements

1. See ANSI Requirements in Chapter 4, Powerline Contact.
2. ANSI B30.22, *Articulating Boom Cranes*, Section 22-1.6.1(c), requires remote-operated cranes to have an emergency stop. Illustration 5 shows a dangerous control arrangement that is accessible to an operator standing on the ground.

Other References

1. See Other References in Chapter 4, Powerline Contact.
2. *Human Engineering Guide to Equipment Design* (U.S. Superintendent of Documents, 1963 (revised 1972)).
3. Wesley Woodson and Donald Conover, *Human Engineering Guide for Equipment Designers* (University of California Press, 1964).
4. Wesley Woodson, *Human Factors Design Handbook* (McGraw-Hill Book Company, 1981).

Suggested Design Criteria

Design controls that are consistent with applicable human factors criteria.

Make remote control systems nonconductive.

Controls for flatbed-mounted cranes should never be operable or reachable from the ground.

Incorporate powerline mapping instructions in operator's manual and point-of-operation warnings as discussed in Chapter 4, Powerline Contact.

Figure 10-2
Remote Control

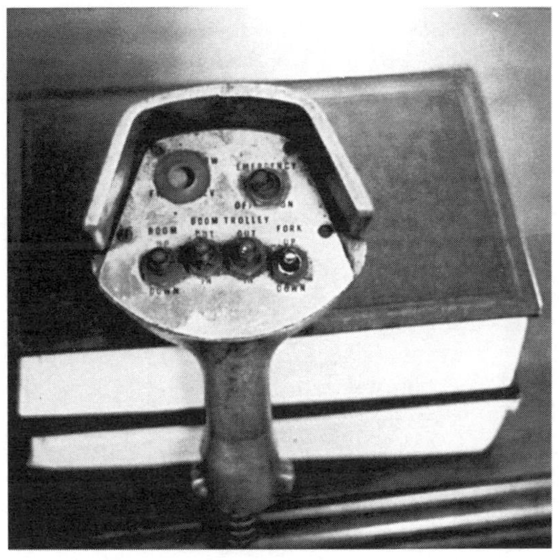

Figure 10-3
Remote Control with Small Levers

Figure 10-4 **Control Accessibility**

Controls accessible while standing on the ground.

Representative Litigation

Plaintiff Tucker; occurred August 23, 1976; 14th Judicial District, Dallas County, Texas.

Operator electrocuted while using a conductive remote control when the boom responded erratically to his command and raised and swung into powerline. Alleged that the conductive control cable was unsafe and that the control system was defective because it did not obey commands given. Case settled during trial.

Unintended Movement of Aerial Lifts

Definition

Aerial lifts and work platforms sometimes drift slightly when raised, making the control levers accessible to activation by outside, intruding objects.

Description

Control levers on some aerial lifts work platforms are unguarded and remain exposed to contact by overhead joists, trusses, beams, pipes, or other building parts when in a raised position. Hydraulic lifts sometimes drift or creep on their own, and this unintended movement can place control levers dangerously close to outside objects where they are vulnerable to activation. (Figures 10-5 and 10-6 show unguarded controls and guarded controls.)

Emergency controls on the ground are also vulnerable to inadvertent activation by other objects. (See previous section: Control Confusion/Inadvertent Activation of Controls.)

Travel controls on self-propelled lifts are also often error-provocative.

Risks Presented by Hazard

Many people who have been working in an aerial lift or platform have been entrapped and crushed when the lift unexpectedly moved.

Available Hazard Prevention Measures

The control panel must be shielded or covered so control levers cannot be activated by an outside object. This also keeps unwanted dirt from falling on it and making control levers inoperable. Another solution is to put a protective cage around the whole work platform so outside objects cannot intrude into it. (See Chapter 23, Maintenance and Renovation.)

OSHA Requirements

OSHA has not commented on this hazard.

ANSI Requirements

1. *Vehicle-Mounted Elevating and Rotating Work Platforms*, ANSI A92.2, Section 4.5.1, states: "Upper controls shall be in or beside the platform, readily accessible to the operator, and protected from damage and inadvertent actuation."

2. *Boom-Supported Elevating Work Platforms*, ANSI A92.5, Section 5.2.3, states: "… Such controls shall be protected against inadvertent operation." Section 5.2.6 states: "All work platforms shall be equipped with an emergency stop device, readily…."

3. *Self-Propelled Elevating Work Platforms*, ANSI A92.6, Section 6.2.5, states: "In addition to the primary operator controls, the work platform shall be equipped with an emergency stop device…." ANSI A92.6 further states that such controls will be protected against inadvertent operation.

Other References

1. *Safety and Health Requirements Manual*, EM 385-1-1 (U.S. Army Corps of Engineers, November 3, 2003), Section 22.J gives additional information on elevating work platforms.

2. *Human Engineering Guide to Equipment Design* by Wesley E. Woodson and Donald W. Conover (University of California Press, 1964), Chapter 8, p. 355, Paragraph 3.2 states: "When the control affects direction of movement of the vehicle, the point of control to which the operator is oriented should move in the same direction as the desired direction of the vehicle."

3. See bibliographic listings 101-107 for other human factors references.

Suggested Design Criteria

The control panel must be protected so levers are not vulnerable to unintended activation by outside objects.

Travel controls should be oriented to reflect the desired direction of vehicle travel.

Figure 10-5 **Unguarded Controls**

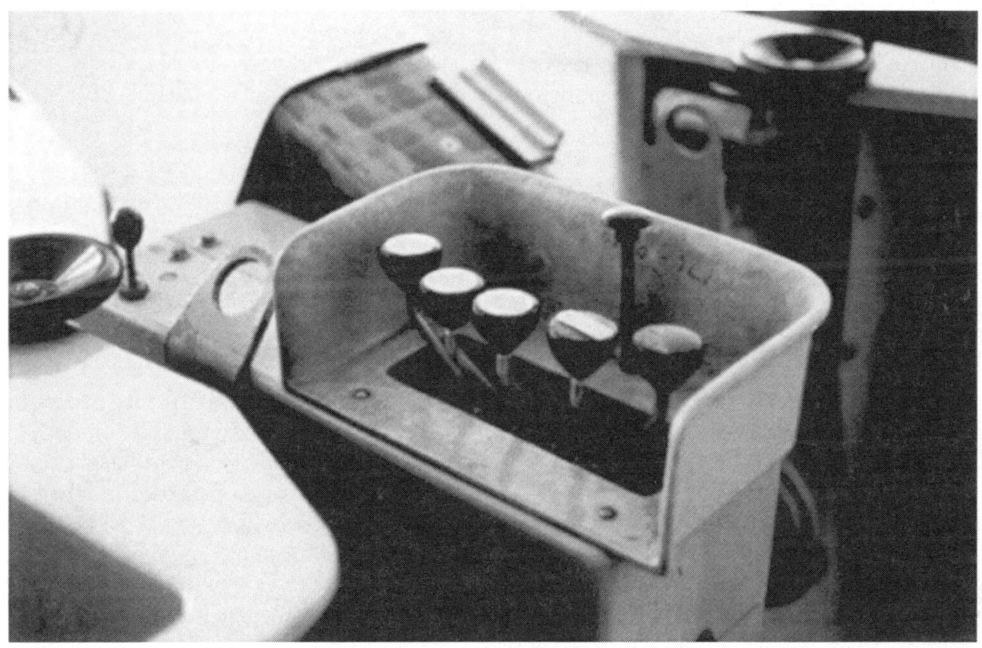

Figure 10-6 **Guarded Controls**

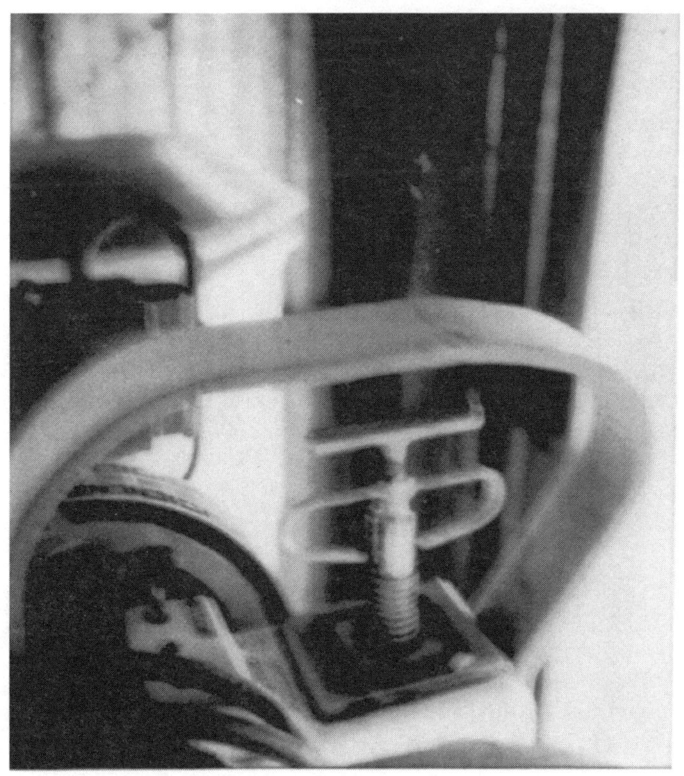

These controls should be separated from the controls for up and down and other movements of the basket to avoid unintentional movement.

Control levers must have boots or covers so dirt cannot accumulate and cause sticking or malfunction.

The hydraulic or other power system should include a drift or creep detector to warn the operator.

A readily accessible panic or emergency stop bar or button needs to be provided so the operator can stop movement to avoid crushing.

A non-conductive cage could be installed that completely encloses the platform to provide a safe work space and prevent crushing by nearby objects due to unintended movement of the lift. (See Chapter 4, Powerline Contact.)

Representative Litigation

Plaintiff Hansell; occurred April 25, 1986; U.S. District, Eastern District, Pennsylvania, #87-5967 and #87-5644.

Deceased worker crushed on aerial work platform between its handrail and a low overhead beam when accumulated dirt from sandblasting caused the control to malfunction. Alleged that besides poor maintenance of controls, the controls should have been covered to make them inaccessible to dirt or other objects. Case settled.

Plaintiff Dubose; occurred June 16, 1986; Beaufort, South Carolina, C/A #2:87-1729-8.

Lineman severely injured when he was in aerial basket and contacted powerlines. Coils of wire stowed in cargo area of truck body became fouled in lower controls, causing bucket to lurch repeatedly against pole. Dead-man switch had been released on operating control handle. Controls were not guarded. Disposition unknown.

Vision

Blind Lifts

Definition

A blind lift is a lift in which the view of the crane operator, rigger, or signaler is blocked so they cannot see one another or in which the site where the load is to be placed cannot be seen.

Description

The operator's view can be obstructed by either the crane's own bulk or by something in the work environment. On mobile cranes traveling backwards, the operator must contend with blind spots on the right side of the crane and to the rear. Crane size can also prevent the rigger, signaler, oiler, and others who are affected by the crane's movement from having direct eye contact with the crane operator. Many other people can be affected by a crane's movement. Welders with hoods on, carpenters, ironworkers, or those engaged in other crafts in the immediate vicinity of crane operations may be outside the range of vision of the crane operator or may be totally unaware that of the crane's movement, as they may be preoccupied with their own tasks and oblivious of other activity. The load may obstruct the view of the operator when making a lift several stories high and make it difficult to see the location where it is to be placed.

Risks Presented by Hazard

The obstructed view of the crane operator and the lack of awareness on the part of other workers as to crane movement contribute to unintentional injury, death, or property damage. In blind lift situations, a simple lift can turn into a life-taking catastrophe almost instantaneously.

Available Hazard Prevention Measures

The key to safe crane operation is the planning of all activities, starting with pre-job conferences and continuing with daily planning to address each new lift situation or change that needs to be made.

The use of an automatic travel alarm is an effective way to give warning to those in the immediate vicinity of crane travel.

To overcome the hazard of lifting loads to or from blind spots, the use of a radio or telephone is much more effective than relying upon several signalers to relay messages by line-of-sight.

A high lift should be considered a blind lift, and an appropriate communication system should be provided.

OSHA Requirements

1. OSHA 1926.16, "Rules of Construction," gives some good guidance for planning even though OSHA only requires it for federally-funded construction projects.

2. OSHA 1910.179(i) states: "Except for floor-operated cranes, a gong or other effective warning signal shall be

provided for each crane equipped with a power traveling mechanism." This also appears in OSHA 1926.550(d)(3).

3. OSHA 1926.201, under a section entitled "Signaling," addresses signals for cranes and refers to appropriate ANSI standards.

ANSI Requirements

1. ANSI has uniform requirements in its various crane standards for hand signals for the various types of cranes (see Illustrations 16a, 16b, and 16c). Most used of these standards for hand signals are ANSI B30.5, *Mobile and Locomotive Cranes,* pp. 36 and 37, and ANSI B30.15-1973, *Mobile Hydraulic Cranes,* pp. 28, 29, and 30.

2. For larger cranes, hand signals are set forth in ANSI B30.2, *Overhead and Gantry Cranes,* p. 28, Figure 5; ANSI B30.3-1975, *Hammerhead Tower Cranes,* p. 18, Figure 4; and ANSI B30.4, *Portal, Tower and Pillar Cranes.* They also address use of communication systems such as telephone, radio, or their equivalent.

Other References

Safety and Health Requirements Manual, EM 385-1-1, U.S. Army Corps of Engineers, published October 1992, Section 16.C.10, requires audio communications when the operator cannot see the load. Also the definition of a "critical lift" on p. 285 includes lifts outside the crane operator's view and requires a critical lift plan as discussed in Section 16.C.17.

Suggested Design Criteria

Install a telephone or radio communication system to allow person-to-person communication during blind or high lifts.

Representative Litigation

Plaintiff Block; occurred November 3, 1988; Circuit, Cook County, Law Division, Illinois, #88-L-20470.

> Injured brain damaged when he fell from a ladder when he was either hit by headache ball or was reaching for descending headache ball. Operator was blindly lowering hook to worker. Al-

leged should have had communication system. Worker should have also been in personnel basket. Case pending.

Plaintiff Brown; Occurred July 17, 1985; 15th Judicial Circuit, Palm Beach, Florida, #86-268(CL)J.

> Injured laborer struck by falling scaffold jack weighing approximately forty pounds that fell when hit by skip during a blind lift. Alleged that communication system was necessary. Case settled.

Plaintiff Ferguson; Filed 1980; 3rd Judicial District, Alaska, #3KN-80-171.

> Oil field worker sustained injuries when knocked from 8-foot-high mud storage tank by a crane moving a hose and not using signaler. Alleged reckless management misconduct in continuing to employ an unsafe crane operator and failing to provide a signaler. Case settled.

Backing

Definition

Backing of a crane is a very hazardous operation because the operator usually has a very limited view to the rear and cannot see if anything or anyone is in the way.

Description

The act of backing a truck-mounted or rough-terrain, mobile hydraulic crane, especially in the pick-and-carry mode, is always difficult because the operator is confronted with a huge blind spot to the rear that makes it impossible to see people or obstructions.

Risks Presented by the Hazard

Because of the operator's limited view to the rear, anyone working at the rear of any crane in this blind zone may not be aware that the crane is being backed.

Backing during pick-and-carry operations is especially dangerous. On some mobile cranes the operator's cab is on the stationary carrier frame and not on the rotating boom pedestal, which severely limits the operator's vision, particularly when backing with a load to the rear.

The operator and signaler may also be preoccupied with controlling a trailing load and may not be always watching the direction of travel.

Available Hazard Prevention Measures

1. Install back-up alarms.
2. Install rear-view mirrors.
3. Use two signalers, one for watching the load and one for backing, in pick-and-carry operations.
4. Install a closed-circuit television.
5. Install near object detection device.

OSHA Requirements

1. OSHA 1910.180(h)(3)(xiv) discusses how a crane will be moved with a load.
2. OSHA 1926.601(b)(4) discusses back-up alarms and signalers.

ANSI Requirements

ANSI has not commented on this hazard.

Other References

Safety and Health Requirements Manual, EM 385-1-1 (U.S. Army Corps of Engineers, October, 1992), Section 16.B.01, requires back-up alarms in addition to signal persons.

Suggested Design Criteria

1. Provide infrared detectors to alert the operator when someone is in the danger zone.
2. Install back-up alarms.
3. Install power-adjusting rear-view mirrors (large West-Coast type).
4. Install closed-circuit television where backing is a repetitive operation.

Representative Litigation

Plaintiff Sweet; occurred May 7, 1984; U.S. Southern District, Florida, #86-6997.

Deceased workman at edge of roadway was struck by an upsetting crane as it was backing down the loop of a freeway interchange under construction. The operator had made a number of successful passes with a load without backing off the shoulder of the roadway. The operator claimed he was unable to see the ground with his mirror. Case settled.

Vision Compromise on Large Track-Mounted Cranes

Definition

Large track-mounted cranes have various configurations. (See Chapter 3, Figures 3-7A and 3-7B.) In many of these cranes, the operator is unable to see pedestrians in the path of intended travel. Some sit in a cab on a tower and cannot see pedestrians immediately below; some sit in a cab mounted on one of the gantry legs and cannot see pedestrians around the base of the other leg because the view is obstructed by the load; and some stand on the floor and use controls mounted on a gantry leg and cannot see if the path of travel is free of pedestrians because this stationary position doesn't allow needed freedom to walk with the control to have unobstructed vision.

Description

Pedestrians are often unaware of crane movement because these large track-mounted cranes can move very silently.

Risks Presented by Hazard

Many people have been crushed or have lost arms or legs when run over by one of these cranes.

Available Hazard Prevention Measures

Equip the crane with a travel alarm to alert pedestrians of movement.

Provide "cow catcher"-type deflectors to brush aside individuals who are in the path of a moving crane.

Provide a panic bar on the fenders to stop crane movement and alert the operator that someone is below.

Provide hand-held moveable controls on a tether or radio controls on floor-operated cranes so the operator can move about to assure the intended path of travel is safe.

OSHA Requirements

1. OSHA 1910.179(b)(6)(ii) states: "Where passageways or walkways are provided, obstructions shall not be placed so that safety of personnel will be jeopardized by movements of the crane."

2. OSHA 1910.179(e)(4), Rail Sweeps, states: "Bridge trucks shall be equipped with sweeps which extend below the top of the rail and project in front of the truck wheels."

3. OSHA 1910.179(h)(4)(i), Warning Device, states: "Except for floor-operated cranes, a gong or other effective warning signal shall be provided for each crane equipped with a power traveling mechanism."

ANSI Requirements

1. *Cranes, Derricks, and Hoists*, ASA B30.2-1943, reaffirmed 1952, Section 1143(a), Truck Fenders, states: Bridge trucks shall be equipped with fenders which extend below the top of the rail and project in front of the truck wheels." Section 1143(b) states: "Trolley trucks should be equipped with similar fenders."

2. *Overhead and Gantry Cranes*, ANSI B30.2.0-1976, Warning Devices, states: "On cab-operated cranes, a gong or other warning device shall be provided for each crane equipped with a power traveling mechanism. On cranes other than cab-operated, a gong or other warning device should be provided for each crane equipped with a power traveling mechanism."

Other References

1. *Safety and Health Requirements Manual*, EM 385-1-1 (U.S. Army Corps of Engineers, November 3, 2003). Section 16.B.02 states: "A warning device or signal person shall be provided where there is danger to persons from moving equipment, swinging loads, buckets, booms, etc."

2. *Industrial Accident Prevention* by David Stewart Beyer, Ph.B. (Houghton Mifflin Company, 1916), on p. 182 states: "All gantry cranes shall be equipped with automatic warning bells."

3. National Safety Council's *Data Sheet 558*, Revision A (1971), Paragraph 8 reads: "The pendant-controlled crane should be equipped with adequate warning alarms. The workman on the floor can shout and possibly be heard more easily than an operator in a cab, but he still needs a warning alarm that operates manually or automatically as required. Therefore, an automatic alarm should be used." And Paragraph 22.n. reads: "A stationary control pendant can create problems and result in unsafe procedures; therefore, the pendant should be on a monorail support running the full length of the bridge. This gives much more maneuverability to the operator, who can move the length of the bridge to observe the lift and, at the same time, stay clear of the lift or any pinch points as the operation proceeds; it also increases the efficiency of the crane."

Suggested Design Criteria

Cranes should be equipped with a travel alarm to alert pedestrians of movement.

"Cow-catcher"-type deflectors should be provided to brush aside individuals who are in the path of a moving crane.

A panic bar should be provided on the fenders to stop crane movement and alert the operator that someone is below.

Hand-held moveable controls on a tether or radio controls on floor-operated cranes should be provided so the operator can move about to assure the intended path of travel is safe.

Representative Litigation

Plaintiff Navarro; occurred early 1970s; Superior County, Maricopa County, Arizona.

Copper smelter laborer lost leg and arm and was thrown into acid-filled trench next to crane track when struck by a moving gantry crane. Operator could not see laborer below and laborer had no warning of crane movement. Alleged that manufacturer failed to provide an automatic travel alarm and cow-catcher and buzzer system on the crane leg accessible to the laborer to alert the crane operator of his presence. Case settled.

Vision Compromise on Straddle Cranes and Crew Cab Failure

Definition

The huge size of straddle cranes limits operators' vision, so they cannot always simultaneously see all four of its legs to determine whether people or other obstructions are in its path.

Description

The operator's station is usually located between the front and rear legs on the left-hand side of a straddle crane. The distance between the front and rear wheels sometimes exceeds thirty feet and sixty feet between each side. These cranes are used to handle pre-stressed concrete beams, ship/truck containers, prefabricated steel components, and other heavy loads. When lifting a load, the operator's view from the left to the right side is obstructed by the load being handled.

Risks Presented by Hazard

Those working adjacent to a straddle crane can be crushed by the wheels of this crane if it unexpectedly begins to travel either forward or backward.

Available Hazard Prevention Measures

To alert the operator that someone is in harm's way, provide a TV monitor so that the area blocked from view by the container being lifted can be seen. Additionally, the use of infrared sensors can warn the operator of people standing next to the wheels.

To warn workers of impending movement, each leg should be equipped with a travel alarm of about 65 decibels that can be easily heard by those close to the leg but does not create an annoyance to other activities.

Additionally, the use of wheel-guard bumpers (cow catchers) as deflectors would push pedestrians harmlessly aside. (See Figure 11-1.)

Use a two-way radio system for person-to-person communication between the operator and ground personnel.

OSHA Requirements

OSHA has no specific requirements for this type of crane but does recommend travel alarms on overhead gantry cranes.

ANSI Requirements

ANSI has no specific requirements for this type of crane.

Other References

Safety and Health Requirements Manual, EM 385-1-1 (U.S. Army Corps of Engineers, October, 1992), Section 16.B.01 covers back-up alarm technology.

Suggested Design Criteria

Provide travel alarms of about 65 decibels and wheel guards for each leg.

Publish safe work procedures in the operator's manual as a guideline for assuring that personnel are in the clear before movement and to limit personnel in the operating area.

Provide an accessory two-way radio system for person-to-person communication between the operator and ground personnel.

Representative Litigation

Plaintiff Braneff; occurred October 20, 1982; 269th Judicial District, Harris County, Texas, #84-17023.

> Injured run over by wheel of straddle crane that was transferring large steel containers from trucks onto railroad cars as he was checking bills of lading for each transfer and was unaware of the crane's movement. Alleged that manufacturer and railroad both failed to provide functional, audible alarms and wheel guards on each leg of the crane. Case settled.

Plaintiff Barnett; occurred January 5, 1977; 17th Judicial Circuit, Broward County, Florida, #78-20171.

> Injured struck in back by backing straddle crane and run over. Alleged that manufacturer failed to provide functional, automatic, audible travel alarm and wheel guards on each leg of the crane. Case settled.

Some straddle cranes are equipped with a crew transport cab, which is suspended just below the operator's cab for easy access from the

Figure 11-1 **Travel Alarms and Wheel Guards for Straddle Cranes**

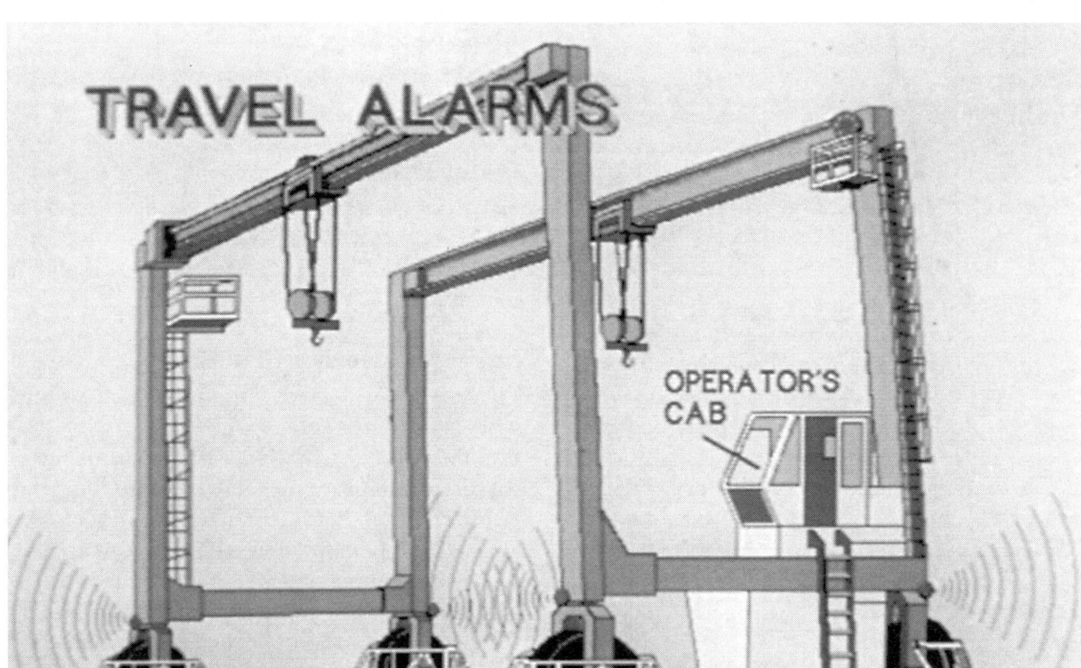

ground. Such a cab needs to be crush-resistant in case the wheel fork becomes dislodged and causes the crane to fall on the cab. The bearings that allow the wheel forks to turn so that the straddle crane can be steered are sometimes designed so that a different size bearing can be inserted. If these are too lightweight, the cause wheel fork can be dislodged. (See Chapter 7, "Operator's Station.")

12

Cable and Sheave Damage

Sheave-Caused Cable Damage

Definition

Wire rope used for hoist line cable can be damaged when sheaves of inadequate diameter or inappropriate groove configuration are used, which also reduces the useful life of the rope. This accelerated wear occurs when the diameter of the sheaves and drums do not meet the optimum diameters recommended by wire rope manufacturers.

Description

Without wire rope, we would not have today's modern cranes. As indispensable as it is, wire rope is very fragile and very vulnerable to bending or pinching as it passes over a sheave or pulley. Sheave and drum diameters should be as large as possible. As a wire rope cable passes and bends over a sheave, the outer strands are stretched and compressive forces are exerted on the inner strands, causing them to be flattened. Wire rope has the ability to stretch in two ways: constructionally and elastically. Constructional stretch is the lay or twist that causes the rope to rotate and lengthen under tension. The elasticity of each separate wire strand that comprises the rope is infinitesimal when compared to that of a rubber band. When a wire strand is stretched beyond its elasticity, it can never return to its former shape and is permanently weakened.

Wire rope can be damaged when the sheave is too small and sharply bends the rope as it passes over it, or when the throat or groove of the sheave is improper. If the sheave throat is too wide to support the wire rope, the rope can flatten; if it is too narrow, the rope will bind. Additional information on how wire rope can fail from foreseeable abuse can be found in rope manufacturers' product manuals. As early as 1947, Roebling published a handbook that provides detailed information concerning how wire rope fails and how safe sheave design can protect wire rope and maximize its life.

Optimum diameters for sheaves and drums recommended by wire rope manufacturers are:

For 6x7 (rope): 42 times rope diameter

For 6x8 (type D): 42 times rope diameter

For 6x19 (rope): 30 times rope diameter

For 6x25 (type B): 30 times rope diameter

For 6x30 (type G): 30 times rope diameter

For 6x37 (rope): 18 times rope diameter

For 8x19 (rope): 21 times rope diameter

For 18x7 (rope): 34 times rope diameter

Safe throat size is assured when the diameter of its groove exceeds the nominal rope diameter. The following chart gives recommended clearances:

Table 12-1 Sheave/Drum Clearances Recommended for Various Nominal Rope Sizes

NOMINAL ROPE DIAMETER (Inches)	RECOMMENDED GROOVE DIAMETER CLEARANCE (Inches)	
	Minimum	Maximum
$\frac{1}{4}$ — $\frac{5}{16}$	$\frac{1}{64}$	$\frac{1}{32}$
$\frac{3}{8}$ — $\frac{3}{4}$	$\frac{1}{32}$	$\frac{1}{16}$
$\frac{13}{16}$ — $1\frac{1}{8}$	$\frac{3}{64}$	$\frac{3}{32}$
$1\frac{3}{16}$ — $1\frac{1}{2}$	$\frac{1}{16}$	$\frac{1}{8}$
$1\frac{9}{16}$ — $2\frac{1}{4}$	$\frac{3}{32}$	$\frac{3}{16}$
$2\frac{5}{16}$ and larger	$\frac{1}{8}$	$\frac{1}{4}$

Sometimes crane specifications for sheaves and drums cannot conform to wire rope manufacturers' optimum requirements because of the crane's configuration for specific uses, and excessive cable wear can occur.

The actual number of lifts and lowerings must also be considered. Just as a car will run out of gas after a certain number of miles are driven, a wire rope will fail after a certain amount of travel over the sheave.

Risks Presented by Hazard

Abuse of wire rope leads to circumstances that can cause the rope to break, become kinked, form a bird-cage, be cut, be flattened, be stretched, high-stranded, or sustain other damage. Bird-caging is caused by torsional unbalance, by forcing a cable through a tight sheave, or by a sudden release of tension. Rope failure creates circumstances that can cause unintentional injury, death, or property damage.

Available Hazard Prevention Measures

Be sure to follow the recommendations of the crane manufacturer for wire rope size, especially when sheaves are of minimal size, which can cause wire rope to wear excessively.

A properly lubricated hoist cable will reduce wear. A poorly lubricated hoist cable can cause excessive wear and dangerous damage to a wire rope. Always use lubricants recommended by the rope manufacturer.

Cable must be inspected daily. Wire rope should be taken out of service when six randomly-distributed, broken wires occur in one lay, or when three broken wires are found in one strand in one lay.

OSHA Requirements

OSHA 1926.550(a)(6) and (7), 1910.179(m)(1), 1910.180(g)(1), 1926.181(g)(1), Marine Terminal Standard 1917.42(b), and Longshoring Standard 1918.63 define and address wire rope requirements.

ANSI Requirements

Inspection, replacement, and maintenance requirements for wire rope are contained in the various ANSI standards covering specific types of cranes.

Other References

1. Wire rope manufacturers have published detailed handbooks on the appropriate application and use of wire rope.
2. The Society of Automotive Engineers' *SAE Recommended Practices*, SAE J881, "Lifting Crane Sheave and Drum Sizes," contains requirements that are not as strict as those of the wire rope manufacturers.
3. "Weighing Your Choices" by Bill Wall (*Craneworks,* January, 1993).

Suggested Design Criteria

Crane manufacturers' operator and maintenance manuals should list which sheaves fall below the cable manufacturer's recommendations and alert crane owners and users of possible accelerated wear and damage to the cable.

For future designs crane manufacturers should reexamine sheave sizes, and when feasible, comply with cable manufacturers' recommendations.

Representative Litigation

Plaintiff Robert Jones; occurred December 6, 1983; Superior Court, Los Angeles County, California, #C525-050.

Worker suffered brain damage when struck by swinging auger attached to crane when securing cable broke as it was drawing the auger up to the crane boom holding position. The crane boom was at an approximate 45 degree angle, causing the unsecured auger to swing in a wide arc. The cable slack was taken up by a hydraulic ram that pulled the cable over the exceedingly small sheaves, causing severe stress and wear on the cable and resulting in premature failure. Case settled.

Cable Kinking

Definition

Wire rope will kink when it is allowed to become slack or is mishandled. It can twist, curl, and double over and sharply bend upon itself.

Description

On cranes, wire rope can kink when slack occurs because of insufficient tension. A kink starts when a loop is formed, is pulled tight, and the natural lay of the rope is destroyed.

Risks Presented by Hazard

A kink in a rope can cause wire rope to break individual strands and ultimately cause the rope to fail under load. For example, on hydraulic cranes, under certain circumstances the hoist line can go slack if it momentarily hangs up on the boom-tip sheave. When the operator attempts to lower the load, the cable can pile up over the hoist drum in loops outside of view. When the wire rope breaks free from the boom-tip sheave, the loops can pull tight into kinks. There are other circumstances in which intermittent failures can cause wire rope on a crane to momentarily go slack into a loop and kink when pulled tight.

Available Hazard Prevention Measures

All wire rope on a crane needs daily inspection.

When wire rope is known to have gone momentarily slack, it should be immediately inspected for kinks and for proper seating in all sheaves.

Fair leads, sheave guides, and a headache ball of sufficient weight should be used to prevent a slack line from binding between the sheave and sheave frame.

Wire rope with kinks must be immediately removed from service and destroyed.

OSHA Requirements

1. OSHA 1926.550(a)(6) and (7), 1910.179(m)(1), 1910.180(g)(1), and 1926.181(g)(1) cover inspection of wire rope.
2. OSHA 1910.179(e)(5) and 1910.179 (h)(1)(ii) and (iii) cover sheave guard requirements.
3. Marine Terminal Standard 1917.42 (b)(2) addresses conditions under which wire rope or wire rope slings should not be used.

ANSI Requirements

Individual ANSI standards for each type of crane include requirements for rope inspection, replacement, and maintenance.

Other References

1. *Safety and Health Requirements Manual*, EM 385-1-1 (U.S. Army Corps of Engineers, November 3, 2003), Section 16.C.17 states: "Whenever a slack line condition occurs, prior to further operations, the proper seating of the rope in the sheaves and on the drum shall be checked prior to further operations."
2. Most wire rope manufacturers have published references on how to avoid kinking and the necessity of removing kinked cable from service.
3. The Society of Automotive Engineers' *SAE Recommended Practices*, SAE J959 (October, 1980), "Lifting Crane, Wire-Rope Strength Factors," contains requirements for wire rope inspection.
4. The National Safety Council's *Accident Prevention Manual for Industrial Operations*, 9th Edition, (1988), discusses wire rope on pp. 134-139.

Suggested Design Criteria

Crane manufacturers should conduct service tests to identify circumstances that allow hoist lines to become slack, determine whether slack cable is caused by binding in sheaves, and, where appropriate, provide sheave guards.

Representative Litigation

Plaintiff Nez; occurred February 10, 1985; District, Albuquerque, New Mexico, #CIV-85-0414-BB.

> Injured became entangled when untangling hoist line that had kinked and piled up on the hoist drum at the base of the hydraulic boom. Alleged that the cable caught in the plastic boom tip sheaves, causing slack in the hoist line between the boom tip and the hoist drum. Case settled.

Hoist Cable Binding in Boom Hoist Tip Sheave on Hydraulic Cranes

Definition

The hoist line can bind between the metal boom-tip sheave-housing and the plastic boom-tip sheave found on some hydraulic cranes.

Description

A plastic sheave can wear more easily than a metal one and can allow slack cable to ride up out of the throat and become momentarily wedged between the sheave itself and its metal housing. The following sequence of events illustrates the active mode of this hazard:

1. The operator lifts a load in a skip above the roof edge.
2. The operator swings and lowers the loaded skip onto the roof.
3. When the operator lifts and swings the skip to clear the roof to return the skip to the ground, slack cable develops. Even though the skip has only been lifted several feet off the roof, the hoist line binds and becomes anchored between the plastic sheave and the sheave frame at the boom tip.
4. The operator thinks he is paying out cable, but all the time the hoist cable is piling up at the hoist drum at the base of the boom, outside of his view.
5. The cable breaks loose with the sound of scraping metal from its bind between the plastic sheave and the metal housing.
6. The skip falls until it comes to the end of the slack.

This type of binding at the boom tip occurs intermittently and may not occur in subsequent lifting. Also, the sheave and boom tip may not outwardly show signs of binding.

Risks Presented by Hazard

This hazard can cause free-fall. If workers are riding in a manbasket, severe injuries or death can result.

Anyone in the vicinity of crane operations may be vulnerable from this type of free-fall.

This hazard can also damage the cable from abrasion and shock-loading.

Available Hazard Prevention Measures

Provide a sheave guard and sheave case that prevents fouling. An anti-two-blocking device prevents damage to the sheave and sheave housing.

Metal sheaves are better choices as plastic sheaves are more vulnerable to wear and damage.

OSHA Requirements

OSHA Subpart N, 1926.550(a)(1) states: "The employer shall comply with the manufacturer's specifications and limitations applicable to the operation of any and all cranes and derricks."

ANSI Requirements

ANSI has not commented on this hazard.

Other References

Wire rope manufacturers provide detailed information concerning the factors which influence sheave and sheave housing design.

Suggested Design Criteria

Re-examine the merits of plastic sheaves over metal sheaves.

Develop sheaves and their housing consistent with wire rope manufacturers' recommendations to avoid binding.

Provide an anti-two-blocking device.

Representative Litigation

Plaintiff Nez; occurred February 10, 1985; District, Albuquerque, New Mexico, #CIV-85-0414-BB.

> Injured became entangled when untangling hoist line kinked and piled up on the hoist drum at the base of the hydraulic boom. Alleged that the cable caught in the plastic boom-tip sheaves, causing slack in the hoist line between the boom tip and the hoist drum. Case settled.

Plaintiffs Beird and McFaddan; occurred June 30, 1989; Court of Common Pleas, Florence County, South Carolina, #90-CP-21-963.

> One worker became paraplegic and several others were injured when raised on skip and hoist line apparently hung up in boom tip sheave. When the operator attempted to lower the skip, the hoist line piled up on the hoist drum at the base of the boom. When the hoist cable pulled free at the boom tip, the piled-up cable rolled over the boom tip sheave, allowing the skip to free fall nearly to the ground, dumping the personnel. Case settled.

Breaking of Boom Hoist Cable Causing Latticework Boom to Fall

Definition

Boom hoist cables on latticework booms can catch between an unusually small boom-hoist sheave and its cover and break, allowing the now unsupported boom to fall.

Description

If a load on a whip line is picked up with a long latticework boom of perhaps 160 feet with a jib of another 30 feet attached, much bounce can be generated that momentarily creates slack in the boom-hoist pendant assembly and allows the boom-hoist cable to catch between the small sheave and its small cover. (See previous section: Hoist Cable Binding in Boom Hoist Tip Sheave on Hydraulic Cranes.)

When lifts are made that require a flat boom, such as breaking form panels loose from cured concrete, a stiff pull is often required. When these form panels break loose, they can cause the boom to bounce. This type of stress places a lot of tension and peening (flattening) on wire rope in small sheaves and can lead to binding and shearing of the boom-hoist cable if it gets caught between the sheave and the sheave cover.

Risks Presented by Hazard

A long, flat boom creates exceptionally high stress on boom-hoist cables, making them vulnerable to failure and causing the boom to fall, endangering those nearby.

Available Hazard Prevention Measures

Avoid lifts with a long flat boom and install sheave guides.

OSHA Requirements

OSHA 1910.179, "Overhead and Gantry Cranes," under (h), Hoisting Equipment, states: "(ii) Sheaves carrying ropes which can be momentarily unloaded shall be provided with close-fitting guards or other suitable devices to guide the rope back into the groove when the load is applied again."

ANSI Requirements

In the Addenda to B30.5 in Section 5-1.7.4(b) and (c), ANSI requires:

> (b) Sheaves carrying ropes which can be momentarily unloaded shall be provided with close-fitting guards or other devices to guide the rope back into the groove when the load is reapplied.

> (c) The sheaves in the lower load block shall be equipped with close-fitting guards that will prevent ropes from becoming fouled when the block is lying on the ground with loose ropes.

Other References

1. SAE J881, Lifting Crane Sheave and Drum Sizes, *SAE Recommended Practices*,

Volume 4 (Society of Automotive Engineers, 1963), allowed for smaller sheaves on cranes, such as boom hoisting sheaves with a minimum of 15 to 1.

2. *Safety and Health Requirements Manual*, EM 385-1-1 (U.S. Army Corps of Engineers, November 3, 2003), Section 16.C.16 states: "Whenever a slack line condition occurs, prior to further operations, the proper seating of the rope in the sheaves and on the drum shall be checked."

Suggested Design Criteria

Design a boom-hoist system with more optimum sheave diameter and include sheave guides for boom-hoist sheaves. Also include in the operator's manual explicit discussions as to limitations of the boom-hoist system when working with a flat boom.

Representative Litigation

Plaintiff Campbell; occurred January 26, 1987; 4th Judicial District, Ada County, Idaho, #90527.

The deceased was decapitated when the jib extension on the boom of a crane being used to remove form panels on a concrete dam under construction dropped on him. The crane's long boom was extended almost horizontally, creating maximum tension on the pin-up assembly that lowers or raises the boom. To dislodge the form panels from the concrete, the hoist line was jerked or reefed, causing the long boom to flex and the pendant system to go slack momentarily. It is also believed that the hoist line may have two-blocked and aggravated this slacking, which caused the boom hoist line to bind in the small sheaves and shear, dropping the boom and jib. Alleged that the sheaves in the boom hoist assembly were too small, that they did not have guides, that no anti-two-blocking was provided, and that the contractor violated reasonable safe crane use by using a flat boom and reefing on the form panels. Case settled.

13

Load Loss

"Killer" Hooks

Definition

Lifting hooks that do not have latches or those with damaged or defective latches that do not retain straps, cables, slings, chains or other rigging safely are sometimes called "killer" hooks. (See Figure 13-1.)

Description

A critical component on a crane when it is lifting a load is a lifting hook that will not let the rigging slip out of the throat of the hook. The most unsafe hook is one without a latch to secure the rigging within its throat. Many hooks have thin, sheet-metal latches that are easily damaged or bent, which fail to safely secure the rigging within the throat of the hook.

Some cranes have load hoist blocks that fit flat against the boom. (See Figures 13-2A and 13-2B.) With this type of hoist block, when the boom is in a lowered position, the rigging rests deep in the throat of the hook, but when the boom is raised to a nearly vertical position, the rigging can slide out of the throat because the swivel can rotate and turn the hook downward.

Normal stress will gradually wear away and enlarge the throat of a hook, increasing the possibility of the rigging's slipping from the hook.

Often a lifting hook is attached to the back of the bucket of a backhoe so rigging can be attached and the machine can be used as a lift-ing device. When the bucket is knuckled under the boom, the throat of the hook is turned downward, allowing the rigging to slip out of the hook.

When two-blocking occurs and the lifting hook is tight against the sheave at the tip of the boom, the lifting hook can rotate up onto the sheave and create a circumstance in which the rigging can become disengaged from the hook.

Even without a load, when there is no latch on a hook, loose rigging can easily be displaced from the hook.

Risks Presented by Hazard

Death, serious crippling injury, or property damage can result when a load is dropped. The use of the thin sheet metal latches is inherently dangerous.

Available Hazard Prevention Measures

Over one hundred positive-type safety latches that prevent hooks from opening have been patented, and the market has available many choices of hooks with safety latches or systems to prevent the rigging from being inadvertent-ly disengaged from the hook. Inherently safe latches are on the swing gate, which require the depression of a latch before the hook will open.

Use of a vertical swivel allows the hook to remain vertical and prevents dumping the rigging from an upended hook.

When lifting is to be done using a backhoe, a ring should be welded to the bucket and a shackle used to attach the rigging.

Wire mousing is available to secure straps within the throat of a hook, but its use in place of a hook latch has proven to be inadvisable. Wire mousing weakens with each use and should not be used more than once.

OSHA Requirements

1. OSHA 1910.180(d)(3)(v) prohibits: "…hooks with cracks or having more than fifteen percent in excess of normal throat opening or more than 10 degree twist…."
2. OSHA 1910.181(j)(2)(ii) for derricks states: "Safety latch type hooks shall be used wherever possible."
3. Marine Terminal Standard 1917.45 (e)(2) states: "Crane hooks shall be latched or otherwise secured to prevent accidental load disengagement."
4. OSHA 1926.550(g)(4)(iv)(B) discusses personnel hoisting in construction and the positive closure of hooks.
5. OSHA 1910.179(j)(2)(iii) and 1910.179 (l)(3)(iii)(a) discusses inspection and maintenance of hooks.

ANSI Requirements

ANSI B30.15, *Mobile Hydraulic Cranes,* under "Load Hooks," Section 15-1.7.6 states: "… Load hooks shall be equipped with safety latches…." and B30.5 of the March 30, 1990, Addenda provides the following:

> 5-1.7.6 Load Hook, Ball Assemblies, and Load Blocks
> Load hook, ball assemblies, and load blocks shall be of sufficient weight to overhaul the line from the highest hook position for boom or boom and jib lengths, and the number of parts of line in use. All hook and ball assemblies and load blocks shall be labeled with their rated capacity and weight. Hooks shall be equipped with latches unless the application makes the use of a latch impractical. When provided, the latch shall bridge the throat opening of the hook for the purpose of retaining slings or other lifting devices under slack conditions (refer to ASME/ANSI B30.10).

Other References

1. "Crane Design Hazard Analysis" by David V. MacCollum, in *Automotive Engineering and Litigation,* (1984), p. 152.
2. Starting with the 9th Edition in 1988 of the National Safety Council's *Accident Prevention Manual for Industrial Operations* in Chapter 4, page 93, it states: "Open hooks shall not be used to support loads that pass over workers or loads where there is danger of relieving the tension on the hook due to the load or hook catching or fouling." NOTE: The reason the 1988 reference is used instead of a current edition is to show how long it has been known that an open hook creates a life-threatening hazard.
3. Approximately 172 patents were issued between 1867 and 1980 that deal with the various types of safety hooks. These patents are listed below:

DATE	NUMBER	INVENTOR	ITEM
02.26.1867	62,309	A. M. Beard	Improvement in Saw Mill Dogs
05.21.1872	126,918	Henry Babcock	Improvement in Couplings for Rope-Bands
09.10.1872	131,224	Viram B. Paul	Improvement in Hooks for Draft-Chains, Whiffletrees, etc.
10.31.1876	183,750	J. C. Clifford	Improvement in Swivels
06.25.1878	205,402	James Marcellus	Improvement in Whiffletree-Hooks
04.15.1879	214,318	Charles Roloson	Improvement in Tackle-Hooks
11.08.1881	249,177	Wm. Pickering	Wire-Rope Attachment
06.16.1885	320,047	P. F. Chambard	Hook
09.14.1886	349,170	T. T. Morrow	Snap-Hook
12.23.1890	443,473	Octave Boiteau	Grab Hook
02.03.1891	445,663	Elmer A. Stiles	Animal-Grip
01.05.1897	574,564	A. L. Nilson	Shackle
02.20.06	812.861	G. W. Martin	Pendent or Trolley-Carrier Meat-Hook

DATE	NUMBER	INVENTOR	ITEM
05.21.07	854,087	Carl Hainlin	Spring Device
07.07.08	892,940	E. O. Davis	Attachment for Crane Hooks
06.29.09	926,156	R. Waterhouse	Hook
08.31.09	932,673	E. A. Benjamin	Snap
03.01.10	950,641	L. C. Sands	Tubing and Casing Hook
03.15.10	952,367	Frank Sherkel	Snap-Hook
08.09.10	967,141	E. B. Merriman	Snap-Hook
05.23.11	992,759	L. S. Denison	Choker Hook
01.23.12	1,015,493	A. S. Howe	Chokes Logging-Hook
05.07.12	1,025,666	W. R. Yeagle	Snap Hook
12.10.12	1,046,795	A. Johnson	Crane Hook Locking Attachment
01.21.13	1,051,114	A. Johnson	Crane Hook Locking Device
05.27.13	1,062,653	C. E. Koons	Snap Hook
08.26.13	1,071,650	J. G. O'Kelly	Safety Hook
09.02.13	1,072,285	R. J. Wigley	Handy Hook
02.24.14	1,088,614	M. T. Olstad	Hook
04.21.14	1,093,630	C. E. Kelly	Snap Hook
11.24.14	1,118,618	E. M. Babb	Draft Hook
06.01.15	1,141,333	H. Hardisty	Checker Hook
06.22.15	1,144,099	L. J. Black	Hoisting Hook
02.22.16	1,173,001	A. P. Keegan	Chain Holding Hook
05.02.16	1,181,258	B. Rosser	Hook
05.16.16	1,183,059	J. O. Balch	Hook
08.08.16	1,193,516	R. Clarke	Safety Hook
10.10.16	1,200,540	E. J. Swedlund	Swivel
02.06.17	1,214,755	D. Cavanaugh	Hook
07.17.17	1,233,376	C. F. Link	Trip Sling-Lock
07.31.17	1,235,480	A. A. Johnson	Snap Hook
09.04.17	1,239,301	H. Pearson	Bull Hook
07.16.18	1,272,705	N. Pearson	Safety Hook
08.13.18	1,275,265	D. Keil	Chain Fastener
10.22.18	1,282,049	C. F. Cummins	Snap Hook
12.31.18	1,289,096	J. Boatright	Safety Hook
12.31.18	1,289,616	K. Berger	Hook
01.14.19	1,291,673	T. N. Robinson	Safety Hook
02.04.19	1,293,543	P. L. Plank	Safety Snap Hook
04.08.19	1,299,424	A. E. Carlton	Hook
04.08.19	1,299,821	J. F. Carpmill	Hook
04.08.19	1,300,231	H. A. Sinclair	Safety Rigging Hook
12.09.19	1,324,676	C. Knudsen	Safety Hook
03.23.20	1,334,830	W. Bastord	Logging Hook
04.13.20	1,336,666	W. G. Vasey	Butt Hook
07.20.20	1,347,369	M. D. Hamrick	Draft Hook
08.24.20	1,350,787	D. J. Darkes	Snap Hook
09.14.20	1,352,982	J. A. Lothi	Snap Hook
10.26.20	1,356,830	G. C. Rohrbach	Safety Locking Device for Hooks
12.07.20	1,361,249	O. L. Giffin	Rope Hook
03.29.21	1,373,235	M. H. Giberson	Safety Hook for Logs
08.02.21	1,386,561	E. B. Foster	Hoisting Device
08.09.21	1,386,894	O. Myrmo	Choker Hook
08.23.21	1,388,494	C. Vala	Butt Hook
09.27.21	1,392,026	R. M. Stevenson	Bull Hook

DATE	NUMBER	INVENTOR	ITEM
09.27.21	1,392,184	B. J. Long	Chain Hook
09.27.21	1,392,260	P. W. Schollar	Retaining and Releasing Device for Ropes, Cordage, and the Like
11.29.21	1,398,887	T. J. Bond	Hook
04.04.22	1,411,549	F. E. Abbott	Guard for Hooks
08.01.22	1,424,238	E. E. Cochran	Butt Hook
10.03.22	1,430,824	A. Martin	Logging Hook
11.14.22	1,435,587	W. H. Crandall	Towing Hook
01.09.23	1,441,378	J. F. Selvidge	Hook
04.10.23	1,451,315	T. M. Avery	Hook Fastener
05.22.23	1,456,264	H. Billmeyer	Safety Snap Hook
06.12.23	1,458,453	F. A. Young	Safety Hook
03.14.23	1,465,804	A. C. Bubb	Coupling Device
11.13.23	1,473,885	E. S. Soule	Hook
11.20.23	1,475,046	J. Bolei	Hook
04.01.24	1,488,744	N. Ekberg	Grabhook for Automobiles
06.24.24	1,498,691	M. J. Kearns	Hook
08.12.24	1,505,051	F. Lindgren	Lock for Hoisting Hooks
09.09.24	1,508,308	F. C. Thompson	Safety Hook
09.16.24	1,508,705	D. E. Mahan	Swivel Hook
10.07.24	1,510,692	A. R. Mielke	Fastener for Anti-skid Chains and other Articles
10.07.24	1,511,002	W. W. Pfautz	Connecting Lock
11.25.24	1,516,806	B. Amrin	Hook
11.25.24	1,517,019	A. A. Serl	Draft Device
11.23.24	1,520,576	J. R. Keaton	Hook
01.06.25	1,521,811	C. W. Hartbauer	Snap Hook
01.13.25	1,522,979	J. P. Ratigan	Safety Connector for Well Works
01.20.25	1,523,765	J. E. Gilchrist	Safety Butt Hook
02.03.25	1,524,699	F. Faber	Safety Clamp
02.03.25	1,524,761	E. Timbs	Well Casing Hook
02.03.25	1,524,844	A. R. Scott	Safety Latch Hook
02.03.25	1,525,292	E. E. Greve	Safety Hook
02.10.25	1,525,753	E. Law	Choker Hook
03.17.25	1,530,010	A. H. Neilson	Safety Hook
04.07.25	1,532,927	T. Nowland	Safety Hook
04.14.25	1,533,995	P. Lang	Hook
04.21.25	1,534,879	M. G. Stewart	Swivel Hook
05.05.25	1,536,617	G. A. Montgomery	Casing Hook for Drilling Apparatus
05.26.25	1,539,551	J. N. Erlandsen	Hook
06.16.25	1,541,991	J. F. Moody	Safety Hook
07.14.25	1,546,208	W. P. Cunningham	Safety Hook
09.22.25	1,554,841	J. Clark	Hoisting Hook
09.29.25	1,555,359	J. Delahunt	Safety Hook
10.20.25	1,557,802	J. Clark Hoisting	Hook
11.03.25	1,559,713	F. C. Lester	Swivel Hook
12.01.25	1,563,341	J. Clark Hoisting	Hook
02.16.26	1,573,444	L. M. Jordan	Hook for Well Drilling Machines
03.09.26	1,576,352	P. A. Nordling	Hook
04.27.26	1,582,345	W. A. O'Bannon	Safety Hook
05.04.26	1,583,347	J. Frischknecht	Snap Hook
08.10.26	1,595,264	I. A. Treiman	Bull Hook

DATE	NUMBER	INVENTOR	ITEM
09.07.26	1,599,087	E. E. Greve	Safety Hook
11.02.26	1,605,799	C. B. Ver Valen	Hook
01.11.27	1,613,641	A. E. Bergquist	Snap Hook
03.29.27	1,622,971	G. N. Porter	Safety Hook
07.19.27	1,636,209	J. Bergsten	Hook
11.22.27	1,650,038	J. E. Potter	Safety Rod Hook
05.01.28	1,667,927	R. Clarke	Safety Hook
05.01.28	1,667,957	E. E. Stevenson	Hoisting Hook
05.15.28	1,669,805	J. P. Beer	Latch Mechanism for Hooks
07.03.28	1,676,167	J. E. Sprain	Safety Hook
08.28.28	1,682,617	W. G. Jensen	Safety Hook
10.16.28	1,688,176	J. Clark	Connector
02.12.29	1,702,087	L. M. Moore	Tug Hook
04.02.29	1,707,721	B. P. Hoffman	Elevator Hook
04.30.29	1,711,346	E. E. Greve	Safety Hook
04.30.29	1,711,440	D. Baker	Hook
05.28.29	1,715,192	D. Fortin	Hook Lock
06.11.29	1,716,997	T. Antoniow	Logging Hook
06.11.29	1,717,129	L. Walasky	Coupling Hook Lock
09.24.29	1,729,188	A. Schillinger	Safety Elevating Hook
02.11.30	1,747,128	W. A. O'Bannon	Tubing Hook
04.08.30	1,753,326	J. C. Yingling	Safety Guard for Hooks and the Like
07.22.30	1,771,314	W. A. Ramsay	Sugar Cane Sling
08.05.30	1,772,390	C. D. Evans	Snap Hook
03.03.31	1,794,694	W. G. Jensen	Safety Hook
07.14.31	1,814,900	C. G. Deppe	Lock Hook
01.26.32	1,842,593	R. Edwards	Safety Hook
03.15.32	1,849,816	J. C. Yingling	Hook Guard
05.24.32	1,860,121	W. C. Trout	Weighted Rod Line Swivel
07.19.32	1,867,574	A. L. Leman	Rod Line Weight
09.27.32	1,879,167	J. B. Freysinger	Safety Snap Hook
09.27.32	1,879,168	J. B. Freysinger	Safety Snap Hook
05.09.33	1,907,436	M. Niemi	Hoisting Block
06.13.33	1,914,189	J. C. Yingling	Hook Guard
03.06.34	1,949,608	E. E. Johnson	Safety Hook
05.01.34	1,956,786	W. J. Bemis	Safety Hook
06.26.34	1,964,428	T. F. Duffy	Safety Hook
07.17.34	1,966,665	W. J. Gourley	G-Link and Lock Therefor
12.25.34	1,985,596	W. C. Burnham	Block Hook
01.14.36	2,027,376	H. E. Grau	Safety Hook
08.24.37	2,091,093	D. Buccicone	Safety Latch for Hooks
08.31.37	2,091,477	H. E. Grau	Hoist Line Hook
11.30.37	2,100,779	R. K. Hertel	Latch for Safety Hooks
05.10.38	2,116,880	H. D. Dee	Self-Locking Cable Hook
06.28.38	2,121,908	E. Ericksen	Hook
05.16.39	2,158,232	H. E. Grau	Drilling Hook
05.16.39	2,158,372	S. W. Long	Spring Hook
01.23.68	3,365,076	L. C. McManus	Device for Stabilizing the Hook Swing of a Crane Boom
04.20.71	3,575,458	E. J. Crook	Hook and Latch with Hook
04.04.72	3,653,102	E. J. Crook	Hook with Gate
07.04.72	3,674,301	E. J. Crook	Hook and Collar with Gate

DATE	NUMBER	INVENTOR	ITEM
03.27.73	3,722,943	John Kalua	Safety Hook
06.26.73	3,741,600	E. J. Crook	Safety Hook
08.06.74	3,827,746	L. C. Byers	Safety Latch for Hoist Hook
10.14.75	3,912,318	D. C. Engh	Self-Closing Lift Hook
03.07.78	4,077,661	Kiichiro Inahashi	Crane Hook Apparatus
04.01.80	4,195,872	C. I. Skaalen	Remote Controlled Safety Hook

Suggested Design Criteria

Manufacturers, after-market suppliers, and crane owners should use only lifting hooks with safety latches. Such hooks should be designed and tested to resist three times the weight of the anticipated load or the crane's maximum lifting capacity.

Representative Litigation

In recent years numerous complaints have been filed where hooks with no retainer latches or the thin sheet metal flapper latches failed and resulted in a serious injury or death. The first of these cases date back to the 1970s.

Plaintiffs Ferguson and Smith; occurred June 13, 1979; Federal Court, Oklahoma, #81-7932-D and #81-792-W.

Two fatalities occurred when lifting a beam to clear building top during first shift. Headache ball two-blocked, curled around boom tip and displaced the straps out of the hook past a defective, thin, sheet metal latch.

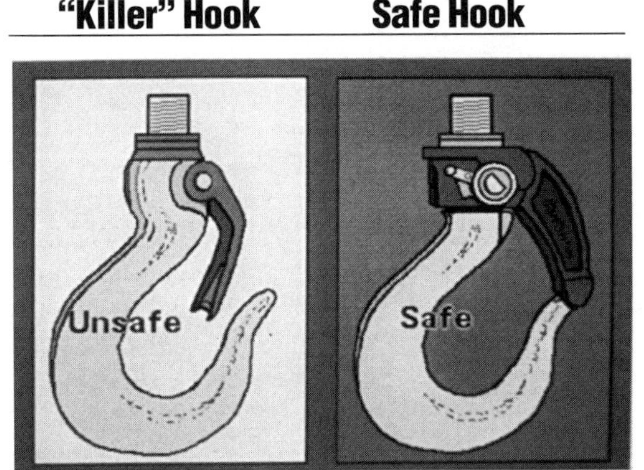

Figure 13-1 **Unsafe Hook/Safe Hook**

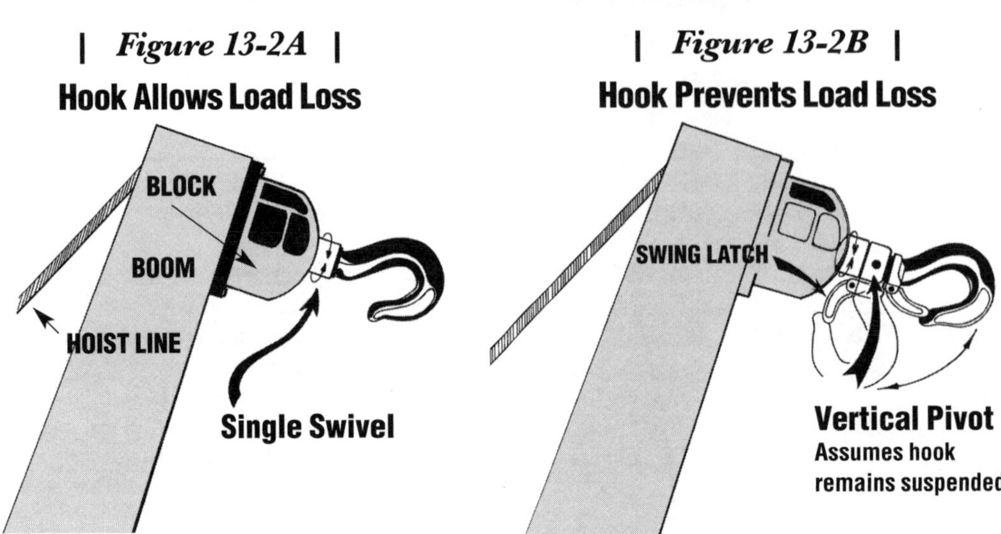

Figure 13-2A
Hook Allows Load Loss

Figure 13-2B
Hook Prevents Load Loss

Plaintiff alleged two hazards: crane should have been equipped with a sturdy, double-latching hook gate and an anti-two-blocking device. Defense verdict, appealed, and settled.

Plaintiff Ellich; occurred approximately 1983; Cook County, Illinois.

Injured struck by falling choker suspended in open, double-throated hook on a derrick. Choker dislodged from hook when it was struck by tip of a boom of another crane that was being slued. Alleged that hook manufacturer also had patents and produced suitable safety latches. Case settled

Plaintiff Selness; occurred May 28, 1997; U.S. District Court, Western District, Tacoma, Washington.

Injured struck by a falling fish tote basket when the strap fell out of the open lifting hook. When the basket brushed against the crane boom, the tension on the strap was relieved, allowing the strap to become disengaged from the open hook, which caused the tote basket to fall. Selness sustained serious permanent injuries.

Free Fall

Definition

Unsuspected or unintentional release of the hoist line that supports the load or the pendants that support the boom can occur in some hoist systems.

Description

Failure of a hoist system that results in dropping either the load or the boom can be caused by the breaking of the hoist line, unintended slack, or unintentional free-wheeling of the hoist drum.

Risks Presented by Hazard

The risk of death or injury is ever present when people are working in the area of boom movement where a load or boom can be dropped upon them. Death and serious injury can also occur when a man-skip falls. Great monetary loss often results when a load containing items of high value is dropped.

Available Hazard Prevention Measures

Select a crane with power-lowering without a free-fall capability and that requires a distinct hand and foot movement to release the brake to prevent free-wheeling of the hoist drum.

OSHA Requirements

OSHA 1926.550(g) and Marine Terminal Standard 1917.45(j)(2) require power lowering when handling manlifts.

ANSI Requirements

ANSI B30.5-1982, *Mobile and Locomotive Cranes*, Section 5-1.6, Controls, states that controls shall be provided with means for holding load hoist, boom hoist, etc. in a neutral position without the use of positive latches. In my opinion, this statement is contrary to sound human factors considerations.

Other References

1. The following references concern avoiding inadvertent control activation:
   a. "Controllers (NI)," *Safety Code for Cranes, Derricks, and Hoists*, ASA B30.2-1943 (The American Society of Mechanical Engineers, 1952), p. 33.
   b. *Factories, The Construction (Lifting Operations) Regulations*, Statutory Instruments No. 1581 (Minister of Labour of England, 1961), p. 8.
   c. "Controls," *Human Engineering Guide for Equipment Designers*, Second Edition (1964), pp. 2-91.
   d. "Pedals," *Military Standard, Human Engineering Design Criteria for Aerospace Systems and Equipment*, MIL-STD-803A-1(USAF), (Department of Defense, January 27, 1964), p. 61.
   e. *Safety Code for Crawler, Locomotive and Truck Cranes*, USAS B30.5-1968, (American National Standard, 1968), pp. 9 and 10,.
   f. *Military Standard, Human Engineering Design Criteria for Military Systems, Equipment and Facilities*, MIL-STD-1472 (Department of Defense, February 9, 1968), pp. 54 and 67.

g. Conover, D. W. and W. E. Woodson, "The Design Implications of Product User Behavior" (The National Commission on Product Safety, Letter Contract No. 70-168, December 30, 1969), p. 125.

h. "Crane-Shovel Basic Operating Control Arrangements," SAE J983, *1971 SAE Handbook, SAE Recommended Practice*, p. 1187.

i. Chapanis, Alphonse, "Design of Controls," *Human Engineering Guide to Equipment Design* (1972), Chapter 8.

j. *Military Standard, Human Engineering Design Criteria for Military Systems, Equipment, and Facilities*, MIL-STD-1472B (Department of Defense, December 31, 1974), pp. 64, 86 and 87.

k. "Controls," *Military Standardization Handbook, Human Factors Engineering Design for Army Materiel*, MIL-HDBK-759 (March 12, 1975), pp. 51 and 52.

l. Woodson, Wesley E., *Human Factors Design Handbook* (1981), pp. 571-643.

m. Huchingson, R. Dale, "Design of Aerospace Controls and Control Systems," *New Horizons for Human Factors in Design* (1981), pp. 145 and 169.

2. The following references concern the human factors aspects of control system design:

a. Woodson, Wesley E. and Donald W. Conover, *Human Engineering Guide for Equipment Designers* (University of California Press, 1964).

b. Meister, David and Gerald F. Rabideau, *Human Factors Evaluation in System Development* (John Wiley & Sons, Inc., 1965).

c. Chapanis, Alphonse, *Man-Machine Engineering* (Brooks/Cole Publishing Company, 1965).

d. Fitts, Paul M. and Michael I. Posner, *Human Performance* (Brooks/Cole Publishing Company, 1967).

e. Surry, Jean, *Industrial Accident Research, A Human Engineering Appraisal* (University of Toronto, Labour Safety Council, May, 1971).

f. Van Cottand, Harold P. and Robert G. Kinkade, *Human Engineering Guide to Equipment Design* (Army-Navy-Air Force Steering Committee, 1972).

g. Huchingson, R. Dale, *New Horizons for Human Factors in Design* (McGraw-Hill Book Company, 1981), See Chapter 5, pp. 145-187.

Suggested Design Criteria

See Chapter 10 on Controls for proper design and location of controls.

Design a hoist so that it can only lower the load under power.

If it is necessary to have a hoist that can drop the load by free-wheeling, separate simultaneous actions by hand and foot should be required to release the brake and the power.

Representative Litigation

Plaintiff Delgadillo; occurred July 9, 1987; U.S. District, Southern District of California, 87-1361-G(M) & Consolidated Cases.

Six people were killed and six seriously injured when riding in a skip being lifted by a large shipyard portal crane that was not moving on its tracks. The operator said he had raised the hoist line and then the boom, losing brake pressure just before he began to slue the crane. Alleged that the control design created an opportunity for inadvertent control activation, resulting in free fall. Case settled.

Plaintiff Peavy; occurred April 14, 1987; U.S. Southern District, Savanna Division, Georgia, #CV487-183.

Deceased struck by free-falling headache ball. Crane operator inadvertently struck control lever when he was hit on head by cab window of crane that fell out while he was adjusting it. Case settled.

Load Loss on Straddle Cranes

Definition
Rough movement or a rough surface can imbalance a load on a straddle crane and cause loss of the load.

Description
When lifting long, bulky, or peculiarly-shaped objects, spreader beams are sometimes used to equalize the load to reduce imbalance. If the lifting hooks do not have safety latches, the load can be inadvertently dropped or fall from one end when traveling on rough ground. This can also happen when the crane has an operator-controlled release system that has no mechanism to retain the load in the hook.

Risks Presented by Hazard
People guiding the load or walking close to it can be crushed when the load falls.

Available Hazard Prevention Measures
Site preparation that includes a level surface for crane travel is extremely important.

Lifting hooks should have sturdy safety latches to prevent inadvertent release, as set forth in Chapter 9, Pinch Points and Nip Points.

Train workers that they should not attempt to guide the load or walk in close proximity of the crane or load during crane travel or load movement. Inform other workers in the area that they should keep out of the way during crane travel or load movement.

OSHA Requirements
OSHA does not address this type of crane.

ANSI Requirements
ANSI does not address this type of crane.

Other References
Refer to operator's and parts manuals for manufacturer's recommended load-holding devices.

Suggested Design Criteria
Include an operator-controlled remote control system for load release. This mechanism should also include a safety latch to prevent unintentional loss of the load during crane travel or load movement.

Representative Litigation
Plaintiff Edwards; occurred October 22, 1974; Circuit Court, Midland County, Michigan, #78-007612-ND-D.

> Flagman injured when load suspended from straddle crane became disengaged from the lifting hook and struck him on the head. Case settled.

14

Boom Failures and Disassembly

Boom Buckling

Definition

When a boom is lowered onto or strikes a structure during sluing, or when a suspended load strikes the boom, the boom cannot sustain such side force, particularly when supporting a suspended load, and can easily collapse. Also, the webbing on latticework boom sections can be bent or broken in shipping or storage.

Description

A boom is very sensitive to buckling not only when subjected to side pull, but also when a crane is making a lift in a very confined location and it strikes a solid structure.

Risks Presented by Hazard

It takes very little force to cause a boom to buckle when it strikes a structure. Those below are in immediate danger of being killed or injured by the falling load or the buckling boom. A damaged boom is a very serious hazard as the boom may collapse or buckle when a lift is being made.

Available Hazard Prevention Measures

It is better to use a tower crane when the lift has to extend over and into other parts of a building.

Use of tag lines will prevent the load from swinging into the boom.

A damaged boom section can only be repaired by the manufacturer or under its direction and supervision.

OSHA Requirements

1. OSHA 1926.550(a)(1) states that the employer shall comply with manufacturers' specifications and limitations.

2. OSHA has not commented on this specific hazard. However, OSHA Subpart C, "General Safety and Health Provisions," 1926.20(a) states: "… that no contractor or subcontractor for any part of the contract work shall require any laborer or mechanic employed in the performance of the contract to work in surroundings or under working conditions which are unsanitary, hazardous, or dangerous to his health or safety." 1926.20(b) states: "… It shall be the responsibility of the employer to initiate and maintain such programs as may be necessary to comply with this part."

ANSI Requirements

ANSI has not commented on this hazard.

Other References

The Society of Automotive Engineers' *SAE Recommended Practices*, SAE J987 (October, 1980), "Crane Structures — Method of Test," and SAE J1063 (October, 1980), "Cantilevered Boom Crane Structures — Method of Test," discuss this hazard.

Suggested Design Criteria

Operator's manuals should contain a warning of this hazard; and maintenance manuals should contain specific requirements for all damaged booms to be repaired by the manufacturer only or under its specific instruction and supervision.

Loss of Stowed Jib Booms on Cantilevered Hydraulic Booms

Definition

A jib boom stored on the side of the main boom can fall when the mechanism for its attachment fails.

Description

The jib is a useful accessory to lift light loads higher or increase the radius of the lift. When not in use, it is stowed and carried on the side of the main section of the boom. For storage, the jib must be totally disconnected from the top extension of the top retractable boom section. It is then rotated 180 degrees so its tip now points down the main boom and is held by a separate support mechanism affixed to the main boom.

Risks Presented by Hazard

If a stored jib falls free, the risk of injury to those working adjacent to the boom is present. Currently, the reliance upon a pin-secured anchor for jib boom storage against the main telescoping boom is the source of foreseeable error. When the pin is improperly placed, the jib boom falls off and either kills or maims someone.

Available Hazard Prevention Measures

Present attachment devices make it possible for those stowing the jib boom to make errors that can result in its coming unattached and falling. A system-safety analysis of jib boom storage attachment devices and the way people use them would be fruitful in determining what should be done to eliminate this hazard. It may be that a slight modification of the present attachment mechanism by the manufacturer would avoid errors by those storing the boom. A spring-loaded latch mechanism that would require manual release might be the answer, eliminating the chance of error on the part of those storing it.

OSHA Requirements

OSHA has not commented on this subject in general but OSHA 1926.550(a)(1) states: "The employer shall comply with the manufacturer's specifications and limitations applicable to the operation of any and all cranes and derricks."

ANSI Requirements

ANSI has not commented on this subject.

Other References

1. The following references on system safety concepts discuss the methodology for identifying this type of hazard:
   a. Wood, Amos L., *The Organization and Utilization of an Aircraft Manufacturer's Air Safety Program* (Fourteenth Annual Meeting of the Institute of Aeronautical Sciences, New York City, January, 1946).
   b. *System Safety, Design Handbooks Series*, 2nd Edition (U. S. Space Administration, 1969 to Present).
   c. MacCollum, David V., "Reliability as a Quantitative Safety Factor," *Journal of the American Society of Safety Engineers* (May, 1969).
   d. Willie Hammer, *Handbook of System & Product Safety* (Prentice-Hall, 1972).
   e. Johnson, W. G. and Grandjean, *The Management Oversight & Risk Tree — MORT* (Superintendent of Documents, 1973).
   f. Johnson, William and Marcel Dekker, *Mort Safety Assurance System* (National Safety Council, 1980).
   g. Willie Hammer, *Product Safety Management and Engineering* (Prentice-Hall, 1980).
   h. Willie Hammer, *Occupational Safety Management and Engineering* (Prentice-Hall, 1981).

Suggested Design Criteria

The manufacturer should make a system safety hazard analysis. One possible solution would be to design and install a spring-loaded, manually-releasing catch that would automatically hold the jib boom securely to the main boom.

Representative Litigation

Plaintiff Gomez; occurred June 21, 1985; 181st Judicial District, Potter County, Texas, #689113.

> Worker injured when stored jib broke loose from boom and landed on him. Suspected that supporting pins were not in place. Case is settled.

Plaintiff Daubert; occurred approximately 1988; Court of Common Pleas, Bucks County, Pennsylvania.

> Rigger injured when storing jib underneath boom. Mast was raised 45 degrees. It was alleged that pin broke and jib swung out, hitting him on head. Outcome unknown.

Latticework Boom Toppling Over Backwards

Definition

When a latticework boom exceeds a 90-degree vertical position, it can easily topple over backwards.

Description

A nearly vertical latticework boom may be toppled by over-pulling on the boom hoist cable or by a hoist line that is two-blocked. Sometimes a crane with its boom in a nearly vertical position will topple over backwards if the crane slues or travels while on a slight slope. High wind can also topple a boom.

Risks Presented by Hazard

This hazard poses a real danger to those not involved in crane operations and who are working behind the crane where they are normally out of harm's way.

Some boom stops are not connected to the boom but are merely positioned next to the boom to resist its backward movement when it reaches a predetermined elevation. Occasionally these stops fail because of misalignment and do not engage the boom.

A construction inspector standing behind a crane was crushed by a toppling boom pulled over by the boom hoist line. When the boom hoist disconnect mechanism, which overrides the boom hoist control lever to prevent the boom from being raised past a safe vertical position, was replaced with a new mechanism so the operator's boom-hoist control could be relocated to conform with ANSI and SAE requirements (see Chapter 10, Controls), the retrofit failed to function because of missing linkage.

The use of wire rope straps to prevent a boom from being toppled over backwards is dangerous as they often fail.

Available Hazard Prevention Measures

In the early 1950s it was recognized that boom stops were essential to avoid booms from toppling over backwards. Boom stops avoid inadvertent over-pulling of either the boom hoist or a two-blocked hoist line. Boom stops do not allow the boom to be raised beyond a safe angle.

There are two types of boom stops. One is a mechanical bumper and the other is an interlock system that prevents further raising of the boom. The mechanical bumper consists of two energy-absorbing cylinders that attach the lower section of the boom to the frame of the cab structure. Some of these energy-absorbing cylinders are spring-loaded devices that offer nominal protection. The most effective barrier-type boom stops have energy-absorbing hydraulic cylinders that stabilize the boom. Hydraulic boom stops also overcome most of the force created by sudden dropping of a load and can aid in overcoming some of the force of unexpected wind gusts. The interlock is generally an electrical or electrical/mechanical system that intercedes and stops the raising of the boom before it reaches a dangerous angle. Many of today's cranes are equipped with both a mechanical bumper and an interlock system to control this dangerous hazard.

An after-market supplier of boom stops is a company called Boom Snub, which manufactures hydraulic boom snubs that function by a bumper action that stabilizes the boom and absorbs any whip action the boom may sustain when hoisting or lowering a load.

At the beginning of each shift, crane operators should determine if the boom-stop system is functioning.

OSHA Requirements

1. OSHA 1910.180(a)(16) defines a boom stop.
2. OSHA 1910.180(h)(3)(xv) states: "A crane with or without load shall not be traveled with the boom so high that it may bounce back over the cab."
3. OSHA 1926.550(a)(17) states: "The employer shall comply with Power Crane and Shovel Association's *Mobile Hydraulic Crane Standard No. 2*," which states:

   7.11.4 Boom Stops — Stops shall be provided to resist the boom from falling backwards on a grade, in a high wind, or in case the hitch fails.

   (1) a fixed or telescoping bumper

   (2) a shock absorbing bumper

   (3) hydraulic boom elevation cylinder(s)

   7.11.5 Boom Hoist Disconnect — A boom hoist disconnect shut-off or hydraulic relief shall be provided to automatically stop the boom hoist when the boom reaches a predetermined high angle.
4. Marine Terminal Standard 1917.45 (f)(2) also addresses boom stops.

ANSI Requirements

B30.5-1982, *Mobile and Locomotive Cranes*, Sections 5-1.9.1(a) and (d) require boom stops and boom hoist disconnects.

Other References

1. *Safety and Health Requirements Manual*, EM 385-1-1 (U.S. Army Corps of Engineers, November 3, 2003), Section 16.D.06 states: "All mobile cranes with cable-supported booms shall be equipped with:

   a. Boom stops which, at the angle specified by the crane manufacturer, limit the movement of that portion of the boom below the point at which the boom stop acts on the boom.

      (1) The boom stop manufacturer shall certify that the boom stop has been designed, manufactured, and functionally tested such that it will fulfill the requirement of SAE J220, Crane Boom Stops (May, 1971). (Pre-1971 cranes will essentially meet the requirement of SAE standard J220 except for Paragraph 4.1.)

      (2) A crane boom stop field test will be conducted to verify the proper setup of the boom stops and functioning of the boom hoist disengaging device. This test will be conducted prior to initiating the load performance test required by Paragraphs 16.C.13. Deficiencies noted shall be corrected prior to the load performance test.

   b. All jibs shall have positive stops to prevent their movement of more than 5 degrees above the straight line of the jib and boom on conventional crane booms.

   c. A properly functioning boom hoist drum shall automatically be restrained from motion in the lowering direction under any rated condition.
2. SAE J220, Crane Boomstop (March, 1991) and SAE J999, Crane Boom Hoist Disengaging Device (February, 1985), *SAE Recommended Practices*, Volume 4, Society of Automotive Engineers.
3. *Crane Handbook* by D. E. Dickie of the Canadian Construction Safety Association of Ontario (1975) gives a good analysis of this requirement on pp. 21 and 24, "Safety Features," and Figures 1.28-1.34.

Suggested Design Criteria

At the time of design, a thorough system safety hazard analysis should be made of both the mechanical-bumper and interlock types of boom-stop systems to assure that the boom cannot be toppled over backwards by any means.

Representative Litigation

Plaintiff Riegel; occurred November 13, 1987; 164th Judicial District; Harris County, Texas, #87-54748.

City building inspector was sitting with others to rear of crane when the boom was pulled over backwards, and he was struck by falling tip of boom. Testimony was given that when operator controls were revised to meet ANSI standards for location, that some of the linkage for the boom hoist stop kickout switch was omitted, and therefore, it failed to stop the boom from being raised beyond its safe, raised position. The operator momentarily left the crane cab and the boom hoist apparently was not completely disengaged, which allowed the boom to continue to raise in the operator's absence. Case settled.

Side Pull

Definition
Side pull is caused by the lateral forces imposed upon a boom when the load line is tensioned to either the right or left of the boom.

Description
When a crane sits on a slope and picks up or suspends a load, the boom is subjected to side pull. Side pull can be operator-induced when attempting to pull a load, hold a load in place, or slue a load. A side pull can arise when two cranes work together to lift a single load.

Risks Presented by Hazard
Side pulls can easily collapse either latticework or hydraulic-cantilevered booms. The longer the boom, the more probable the hazard of boom collapse will occur.

Available Hazard Prevention Measures
Cranes must be made level before use. Leveling bubbles or other sensing systems are available to aid the operator.

Sensing devices can also be installed that inform the crane operator that a side pull is developing.

OSHA Requirements
1. OSHA 1910.179(n)(3)(iv) states:

"Cranes shall not be used for side pulls except when specifically authorized by a responsible person who has determined that the stability of the crane is not thereby endangered and that various parts of the crane will not be over-stressed."
2. OSHA 1910.180(h)(3)(iv) states: "Side loading of booms shall be limited to freely suspended loads. Cranes shall not be used for dragging loads sideways."
3. Marine Terminal Standard 1917.45(d) and Longshoring Standard 1918.74 (a)(7) states that equipment shall not be used in a manner that exerts side-loading stresses.

ANSI Requirements
All ANSI standards for various types of cranes state that side loading of booms shall be limited to freely suspended loads, and that cranes shall not be used for dragging loads sideways.

Other References
1. *Safety and Health Requirements Manual*, EM 385-1-1 (U.S. Army Corps of Engineers, November 3, 2003), Section 16.D.02 states: "All lattice boom and hydraulic mobile cranes shall be equipped with the following: ... b. A means for the crane operator to visually determine the levelness of the crane.
2. *Crane Handbook* by D. E. Dickie of the Canadian Construction Safety Association of Ontario (1975), discusses side pull on p. 108, and illustrates side pull in Figures 4.25 and 4.31.

Suggested Design Criteria
Install a bubble indicator to assist the operator in leveling the crane. In addition, an electronic alarm is needed to advise the operator that a change of circumstances has taken place and the crane is no longer level.

Representative Litigation
Plaintiff Hubble; occurred approximately 1975; Los Angeles, California.

Deceased decapitated when boom collapsed from a side pull when working with another crane in moving an oil derrick. Case settled.

Plaintiff Fields; occurred October 29, 1976; Federal Court, U.S. Eastern District, Oklahoma, #77-74-C.

Deceased workman struck by collapsing crane boom used to lift 235,000-pound railroad locomotive back onto tracks, step by step, one end at a time. Alleged that this sluing action created side pull. Case settled.

Improper Boom Disassembly on Latticework Boom Cranes

Definition

An unsupported, suspended boom can fall upon those removing the pins under it. (See Figure 14-1 showing unsafe boom disassembly.)

Description

Latticework booms are disassembled for shortening, lengthening, or transporting. When the boom is lowered to a horizontal position and suspended from the boom tip by its pendant guys, boom collapse can occur if each boom section is not individually supported at both ends by blocking or cribbing. This can also happen when the tip rests on the ground and the sections are not supported. If the lower pins that connect the boom sections are the first ones knocked out by workers who are under the boom, the boom can collapse downward upon them.

Risks Presented by Hazard

A recent tabulation of crippling injuries/deaths resulting from this hazard shows some 100 documented occurrences. Improper disassembly can cause a latticework boom to collapse and is a serious, life-threatening hazard that has caused many deaths and crippling injuries. Three primary circumstances lead to injury when latticework boom sections are being disassembled:

1. Cribbing is not used to support each boom section at both ends.
2. Workers are unfamiliar with the equipment.
3. A poor location is chosen for dismantling.

4. A time limit is set in which to complete the function to meet the task deadline.

In my view, incorrect disassembly of crane booms can often be traced to manufacturers' instructions that discuss procedures for lowering the boom tip to the ground and relocating the pendant lines to a new location to provide support but do not reference the need to have cribbing in place before removing connecting pins. Such incomplete instructions have led to death or serious crippling injury and are basically in violation of requirements that prohibit work under suspended loads or unsupported equipment.

Available Hazard Prevention Measures

Provide written procedures for boom disassembly that include blocking or cribbing for each boom section.

Post a warning label that shows safe disassembly procedures on the crane cab as illustrated in Figure 14-2. Also post a warning label on all boom sections showing the danger, as illustrated in Figure 14-3.

Train employees to use blocking or cribbing as set forth in the written procedures.

Immediately prior to boom disassembly, conduct a review of written procedures with the crew and assure cribbing materials are made available and in place before any pins are removed.

When reassembling a boom, insert all pins from the inside out so the head is on the inside. During the next disassembly, this will allow the pins to be driven from the outside in and limits worker exposure to any unexpected shifting of the boom section on the cribbing when the pin is removed. Also, the same objective can be achieved by using double-ended pins supplied by the crane manufacturer. (See *Crane Handbook* by D. E. Dickie, Canadian Construction Safety Association of Ontario (1975), p. 74, Figure 3.34 and p. 78, Figure 3.41.)

Crane boom sections that are bolted together must be supported by cribbing before disassembly is commenced. In the past, during disassembly it was found that workers sometimes used cribbing when boom sections were connected by bolts parallel to the length of the boom because as the bolts would bind unless the boom sections were fully supported by cribbing.

OSHA Requirements

1. OSHA 1926.600(a)(3)(i) states: "Heavy machinery, equipment, or parts thereof, which are suspended or held aloft by use of slings, hoists, or jacks shall be substantially blocked or cribbed to prevent falling or shifting before employees are permitted to work under or between them."

2. OSHA 1910.179(n)(3)(vi) states: "The employer shall require that the operator avoid carrying loads over people." A suspended boom that is not cribbed is not safe to disassemble.

3. OSHA 1910.179(n)(3)(x) states: "The employer shall insure that the operator does not leave his position at the controls while the load is suspended." This assures that the operator does not leave the cab during disassembly.

4. OSHA 1910.180(h)(3)(vi) and OSHA 1910.180(h)(4) contain the same requirements as above for crawler locomotive and truck cranes.

5. OSHA 1926.550(a), General Requirements, relies upon the manufacturer's instructions on how to safely disassemble the boom.

6. Marine Terminal Standard 1917.45 (f)(12) discusses assembly and disassembly of crane boom sections.

ANSI Requirements

ANSI standards in general have prohibited work under unsupported, suspended loads.

Other References

1. *Safety and Health Requirements Manual*, EM 385-1-1 (U.S. Army Corps of Engineers, November 3, 2003), Section 16.D.08 states:

   a. The manufacturer's boom assembly and disassembly procedures shall be followed. The manufacturer's boom assembly and disassembly procedures shall be reviewed by all members of the assembly/disassembly team prior to assembly and disassembly.

   b. When removing pins or bolts from a boom, workers shall stay out from under the boom. Sections shall be blocked or otherwise secured to prevent them from falling, when necessary.

2. *Crane Handbook* by D. E. Dickie of the Canadian Construction Safety Association of Ontario (1975) has detailed discussions on boom disassembly. See Figures 3.33, 3.34, 3.36, 3.37, 3.38, 3.39, 3.40, and 3.41 on pp. 73-78, with special discussion on pp. 65 and 66.

3. Operators' manuals for various cranes usually have instructions on boom disassembly, but often fail to specify the use of cribbing.

Suggested Design Criteria

Operator's manuals need to specify the use of cribbing or other support systems for each boom section, so when pins or bolts are removed, the sections cannot fall.

Design should be developed that provides automatic full support on each boom section so that disassembly can be accomplished without unintended collapse.

D. E. Dickie in his *Crane Handbook* describes the use of step pins that can only be inserted from inside facing out and that can only be removed by driving them from the outside in (see Figure 3.39 on p. 78) and the use of welded lugs on the outside that prevent pins from being inserted from the outside in and than can only be removed by driving them from the outside in (see Figure 3.40 on page 78).

A label should be provided on the crane cab to show how the boom can be disassembled safely when supported by cribbing and on the boom at each connection to warn of the danger of collapse when unsupported.

Representative Litigation

Available data shows that unsafe boom disassembly has caused at least sixty deaths, two paraplegias, and many serious injuries have been litigated. Many boom collapse cases have been settled with the records sealed by the court at the request of defense attorneys, allowing the hazard to continue to kill and maim by restricting the transfer of important injury data and available hazard prevention information. Safety professionals need this information to inform management and workers of hazards that not only take lives but are exceedingly costly.

Plaintiff Allbritton; occurred August 10, 1983; Circuit Court, Cook County, Chicago, Illinois, #73-L-11715.

Deceased was removing pins from an unsupported boom. Boom collapsed when pins were knocked out. Alleged that pin system for connecting boom sections was unsafe. Verdict for plaintiff.

Plaintiff Hedgepath; occurred June 29, 1977; U.S. Southern Division, Hattiesburg Division, Mississippi, #H81-1086(N).

Injured assisting in disassembling boom to add additional section when boom collapsed and crushed him when pins were knocked out. He is now a paraplegic. Alleged that pin system for connecting boom sections was unsafe. Case settled.

Plaintiff Johnson (Mauseth); occurred July 12, 1979; District Court, Clay County, Minnesota, #C-81-178.

Deceased crane operator crushed when he disconnected lower boom pins without reconnecting pendant lines and boom collapsed on him. Alleged that pin system for connecting boom sections was unsafe. Case settled.

Plaintiff Underwood; occurred March 17, 1994; U.S. District Court, Western District of New York, Buffalo, New York, #96CV0855.

Injured instructed to dismantle the crane boom, believed the boom snubs were supporting the boom, and was unaware of the danger. Lack of warnings and manual provided no instructions as to the need to hitch pennants to prevent collapse.

| *Figure 14-1* **Unsafe Boom Disassembly** |

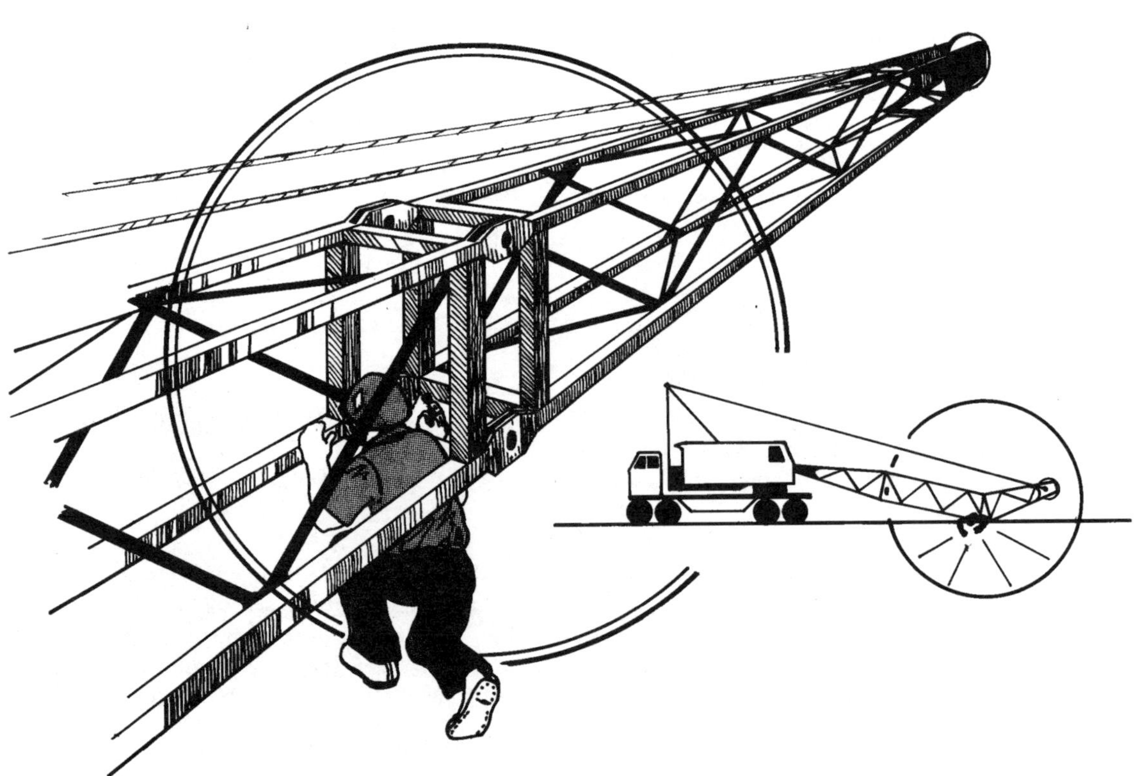

| *Figure 14-2* **Boom Disassembly Label for Crane Cab** |

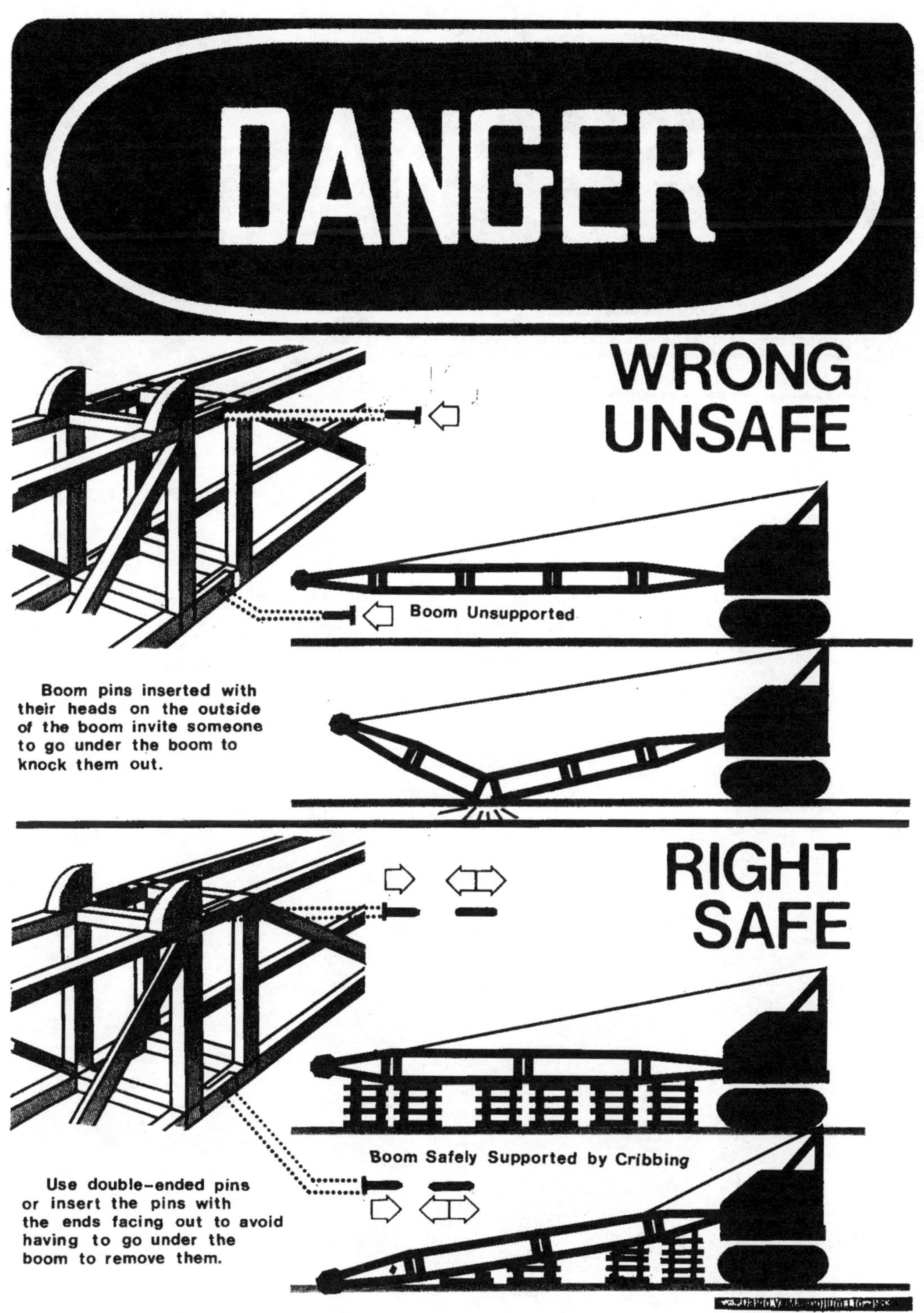

Figure 14-3 Boom Disassembly Label for Crane Boom

15

Transportation

Bouncing Booms on Cranes Being Transported

Definition
If a boom on a mobile crane is unsecured while being transported on a flatbed truck or trailer, it can bounce up and down several inches above its normal height.

Description
When a wheeled crane is first loaded onto a truck or trailer, the combined height of the truck or trailer and the crane may not appear to impede passage under highway underpasses, but pavement irregularities can cause strong vibrations and magnify the normal bounce of the tires or an unsecured boom. This vertical movement can range from inches to a foot or more.

Risks Presented by Hazard
During highway travel, if the tip of the boom has not been secured and is allowed to bounce up and down, it can strike the support beams of a highway underpass, causing severe damage to the underpass. It can also dump the crane from the truck or trailer, endangering vehicles traveling behind.

Available Hazard Prevention Measures
Securely anchor the boom to prevent movement during travel.

Develop procedures for transporting mobile cranes.

Provide adequate training that addresses proper anchoring to prevent boom bounce during transport on flatbed trucks or trailers.

OSHA Requirements
OSHA has not commented on this hazard.

ANSI Requirements
ANSI has not commented on this hazard.

Suggested Design Criteria
Provide a boom anchor system.

Provide detailed procedures for transport in the operator's manual.

Representative Litigation
Plaintiff Plattor; occurred November, 1978; District Court, El Paso, TX, #79-5921.

> Injured was driving behind a truck carrying a crane. The crane boom bounced while passing under an underpass, striking the center beam and dumping the crane onto roadway. The underpass had to be replaced. Case settled.

Adjusting A-Frames on Latticework-Boom Cranes to Accommodate Low Underpasses

Definition
Roads and highways often have underpasses with low 14-foot clearances. Assembled latticework crawler cranes often have an A-frame clearance of more than 14 feet.

Description
A-frame clearances that narrowly prevent passage through underpasses create strong incentives for crane crews to temporarily lower the A-frame by using straps or their own modifications to avoid having to disassemble the entire boom to move the crane to the next job site.

Risks Presented by Hazard
Serious injury can be sustained by personnel attempting to lower or raise the A-frame while it is supporting the weight of the extended flat boom.

Available Hazard Prevention Measures
Select a crane that has an A-frame low enough to allow passage under most highway underpasses.

OSHA Requirements
OSHA has not commented on this hazard.

ANSI Requirements
ANSI has not commented on this hazard.

Other References
Human factors textbooks cover the concept that good design meets foreseeable operator and user needs. (See bibliographic listings 101-107.)

Suggested Design Criteria
Provide a lower A-frame that can go under 14-foot underpasses, or design one that can be mechanically lowered.

Representative Litigation
Plaintiff Andrews, occurred approximately 1978; Circuit Court, Cook County, Illinois, #78-L-17982.

The injured was struck by the A-frame mast that supports the boom hoist cables. The boom and mast were lowered to go under an underpass. After passing through the underpass, the temporary, makeshift rigging failed when a pin broke. Testimony revealed that a retired crane operator had previously advised the crane manufacturer's sales engineer that a simple design modification was needed to afford safe lowering of A-frame to go under low underpasses. His safety suggestions went unheeded. Case settled.

Absence of Communication Between Carrier Cab and Crane Cab

Definition
When a mobile crane is in the travel mode, a need arises for direct voice communication between the crane operator and the carrier driver, particularly in circumstances where an extended crane boom must be slued to accommodate corners, curves, intersections, buildings, or other obstructions.

Description
A latticework boom is often trailed behind a mobile crane to avoid having to change its length for travel. The crane operator can only see to the rear, so needs the carrier driver's advice as to where the boom needs to be positioned to accommodate road conditions. In many of these traveling circumstances, a signal person rides in a second vehicle, requiring radio communication among all three individuals.

Risks Presented by Hazard
Lack of communication can cause death and injury, not only to those maneuvering the crane but to those who are also using the roadway.

Available Hazard Prevention Measures
Provide a person-to-person phone or a radio communication system between the carrier driver and the operator in the cab. Where other guide vehicles are used to convoy mobile cranes, all parties need to be in radio communication.

Provide a seat belt for the crane operator's use when the mobile crane is in the travel mode. (Also see Chapter 7, Operator's Station.)

OSHA Requirements

OSHA has not commented on this hazard.

ANSI Requirements

ANSI has not commented on this hazard.

Other References

Human factors text books have information concerning the need for person-to-person communication. (See bibliographic listings 101-107.)

Suggested Design Criteria

Provide a person-to-person communication system between the crane operator and the carrier driver.

Provide a seat belt in the cab for the crane operator. (See Chapter 7, Operator's Station.)

Representative Litigation

Plaintiff Spaniola; occurred June 16, 1987; 17th Judicial Circuit, Broward County, Florida, #88-28470-CO.

The injured sustained severe brain damage when he apparently fell through an open crane cab door, or was ejected, while the crane was being driven around a curve to exit a freeway. Alleged that the lack of a communication system, seat belts, and ventilation or air conditioning system so the door could be closed during travel in hot weather caused this injury. Case settled.

Loading and Unloading Cranes From Trailers and Trucks

Definition

Cranes are often overturned during the loading and unloading process.

Description

If proper accessories or facilities are not used and if specific procedures are not followed, the moving of a wheeled or crawler crane on and off an elevated trailer or truck bed can be a difficult process.

Trailers and trucks used to carry cranes are very susceptible to sinking or becoming stuck on temporary access roads due to the weight of the crane. If this happens, the crane needs to be off-loaded so the trailer or truck can be pulled onto stable ground.

Risks Presented by Hazard

When moving a wheeled or crawler crane off a tipped trailer or truck bed, the likelihood of its slipping sideways and overturning is great.

Because cranes have a high center of gravity, they can be vulnerable to overturning while being moved up or down ramps if they become even slightly off center or if they are on a slippery surface.

Available Hazard Prevention Measures

Confirm that the truck or trailer is adequate for transporting the crane. A wooden bed is often safer than a steel one as it is not as slippery. Side curbs or center curbs often prevent sideways slipping of the crane as it is loaded or unloaded and will keep it on track.

Assure that the loading and unloading location and travel route can support the weight of the crane and the transporting vehicle.

OSHA Requirements

OSHA has not commented on this subject.

ANSI Requirements

ANSI has not commented on this subject.

Other References

The Society of Automotive Engineers' (SAE) Handbook for On-Highway and Off-Highway Machinery contains references to trucks and trailers but does not address proper design to assure for safe loading and unloading of large equipment.

Suggested Design Criteria

Operator's manuals should contain information on safe crane transport by truck or trailer, should define the ideal type of transporting equipment, and should give detailed instructions on safe loading and unloading, as well as methods of anchoring.

Representative Litigation

Plaintiff Duett; Occurred July 24, 1991; Circuit, Humphrey County, MS, #92-4836.
>   Deceased drowned when excavator slipped sideways off trailer, stuck in mud of a pond, and overturned. Case settled.

16

Counterweights

Dropping of Removable Counterweights and Extendible Counterweight Pinch Point

Definition

Many cranes have removable or adjustable counterweights to make the crane easier to transport. Methods used to remove or adjust these counterweights are often error-provocative and dangerous.

Description

The need to add more counterweights increases as cranes are designed to lift greater loads higher and farther. Because many of our streets, highways, and bridges have load limits, many cranes arrive at the work site without counterweights attached.

Counterweight systems are designed in two ways. One features detachable weights, and the other makes the counterweight extendible to the rear to achieve balance. Most cranes rely upon another crane to load and unload counterweights. Some cranes have a self-lifting system to load and unload the counterweights onto a trailer. Some of these self-loading systems can only lower the counterweight to the level of the trailer and cannot place or pick up the counterweight from the ground. In my view, this design which invites dangerous innovation in the event a trailer is not available.

A few cranes with counterweights located about five feet above the ground use hydraulic rams to extend the counterweights. Some have storage compartments under the counterweight. In one instance, a worker's head was crushed as he was putting an outrigger pad into one of the storage compartments under the counterweight when the extended counterweight was being retracted. The controls for the hydraulic ram were located where the individual operating them could not see that someone was standing between the extended counterweight rams.

Risks Presented by Hazard

The dropping of a counterweight has caused serious injury. Extendible counterweights can be error-provocative.

Available Hazard Prevention Measures

Crane crews should be trained to follow carefully the manufacturer's step-by-step instructions for the handling of counterweights.

Those who supervise crane operations should closely monitor the handling of counterweights to assure that this task is being performed in a manner consistent with the manufacturer's instructions.

OSHA Requirements

OSHA Subpart C, "General Safety and Health Provisions," 1926.20(a) and (b), require a safe workplace.

ANSI Requirements

ANSI has not commented on such systems.

Other References

See manufacturers' operator's manuals for crane being used.

Suggested Design Criteria

System-safety fault-tree hazard analyses should be conducted when designing counterweight systems to identify hazards that may be encountered and errors that operating personnel may make. Appropriate hazard prevention features should be provided.

Extendible counterweight systems should have 7 ½-foot clearance above the ground to eliminate this pinch point. (See Chapter 9.) Storage compartments should be removed from the counterweight area.

Operators' manuals should give step-by-step procedures for the safe handling of counterweight systems. These same instructions need to be included on labels mounted on the outside of the crane cab so they can be readily seen by operating personnel.

Representative Litigation

Plaintiffs Dorman and Donovan; occurred March, 1978; District, Jefferson County, Texas, #D109,669.

> Injured lost both legs at his groin when a counterweight fell on him. The crane was designed with self-loading counterweights that could only be lifted when on its carrying trailer. The counterweight had been placed on the ground so the trailer could be used elsewhere. The counterweight fell when they attempted to raise it from the ground and into place using another crane. Alleged that lifting device should have been able to lift the counterweight from the ground. Settled during trial.

Plaintiff Beresford; occurred November 18, 1987; Dallas County, Texas.

> Deceased's head was instantly crushed while he was placing an outrigger pad in its storage compartment underneath the cab's counterweight that was extended on hydraulic rams. The oiler did not see the deceased and retracted the counterweight. Alleged that controls should be located so whoever uses them can see the area of counterweight movement and that the storage compartment was in an unsafe location. Case settled.

Upset of Counterweight-Stabilized Aerial Lifts

Definition

Self-propelled aerial lifts with counterweights do not have outriggers as the counterweight provides stability. This type of lift can upset backwards on a slope if its mast is in a vertical position because the weight of the mast and the weight of the counterweight are now on the downhill side.

Description

Lifts sometimes upset over backwards on a slope when the operator working aloft is unaware that the tires have sunk into the ground and unbalanced the lift or when the operator may not have perceived that the lift was parked on a slope. In either of these situations, if the boom is rotated while the operator is aloft, the counterweight is then suddenly positioned on the downhill side of the slope, causing the raised lift to topple over.

Risks Presented by Hazard

When backward instability occurs, anyone in the lift can fall many feet to the ground, depending upon the reach of the lift. If the raised mast of the lift happens to fall against a structure, it is often very difficult to rescue anyone in the lift.

Available Hazard Prevention Measures

Audible slope indicators are available to warn the operator of a potential danger.

OSHA Requirements

OSHA relies upon ANSI A92.2, *Vehicle-Mounted Elevating and Rotating Aerial Devices.*

ANSI Requirements

Boom-Supported Elevating Work Platforms, A92.5, Section 6, "Warning Devices," states: "All work platforms shall be fitted with an alarm or other suitable warning at the platform, which will be activated automatically when the machine base is more than 5 degrees out of level in any direction."

Other References

Bibliographic references 125-133 on system safety provide helpful guidance on how to conduct a fault-tree analysis that would reject error-provocative design.

Suggested Design Criteria

Include audible slope indicators in design.

Install a load-moment limiter (LML) to intercede and prevent movement of the boom in any configuration that would cause upset. (See Chapter 5, Upset.)

Incorporate into the operator's manual a specific discussion about the relationship of soil strengths and soil-supportive capabilities to tire size. (See Chapter 5, Upset.)

Representative Litigation

Plaintiff Burlingame; occurred October 19, 1981; U.S. Western District, Northern Division, Michigan, #84CV9410BC.

> Injured had both arms crushed when self-propelled aerial work platform with a nearly vertical mast tipped part way into a tall bridge pier. He was working from the platform filling in form anchor holes in the pier with concrete, with the platform positioned towards the center of the pier on a slight slope. In order not to have to move the aerial work platform to fill in the holes on the other end of this pier, he decided to rotate the mast approximately 180 degrees. When the mast had been rotated about 120 degrees, the lift toppled over and fell into the pier's other side, luckily not going all the way over. Alleged that the manufacturer failed to provide a control system that would prevent operation of the boom if the machine was not on a level surface. Case settled.

17

Turntable Undocking and Boom Hinge Pin Assembly Failure

Definition

The turntable on a crane can fail, causing the cab, boom, and counterweight to fall from the carrier (undock). The boom hinge pin on hydraulic cranes can become disengaged and cause the boom to fall off.

Description

During a heavy lift, the bolts holding the crane hoist structure to the turntable ring or securing the pedestal have been known to fail. An overload usually triggers such failure. Investigation usually reveals that the connecting bolts or pins had become loose and/or were fatigued from previous over-stressing.

Unintentional overloading can occur and over-stress the turntable assembly if the operator relies upon shipping weights, which may or may not be accurate. The boom hinge pin assembly is required to sustain a variety of forces that occur when raising, lowering, sluing, extending or retracting the boom. If the boom hinge pin is not properly secured, it can disengage.

Risks Presented by Hazard

Because cranes are designed to operate under such close tolerances, a lift in excess of rated capacity is possible, particularly when the lift is over the cab of the crane's carrier frame. (See Chapter 5, Upset.)

Investigation of turntable failure revealed that crane operators are generally unaware of this hazard. Serious consequences can occur when the boom undocks or disconnects at its hinge point, causing the boom to fall and endanger those working with the crane or guiding the load.

Available Hazard Prevention Measures

Inspection of the turntable assembly should be part of the annual inspection to determine whether damage or dangerous over-stressing has occurred. Installation of load-measuring devices would assist operators in avoiding overloading. Reliance on friction to hold bolts in place does not prevent them from loosening as vibration and structural stresses occur. Critical bolts used to secure hinge pins as a locking device need to be designed so that back systems are free of stress or the effects of vibration. One solution is to use castle nuts using a drilled bolt-hole with a cotter key, lock or spring-type washers, or anaerobic adhesives that harden after assembly. When in doubt, the inspection and analysis of critical assemblies should be made by a licensed professional engineer who has specialized in fastening systems and who can analyze the manufacturer's design documentation. Designers should define in operating manuals the specific torque to which critical bolts must be tightened and provide guidance as to the type of bolts so that ones with proper strength are used as replacements.

OSHA Requirements

OSHA 1910.180(h)(3)(vii) states: "On truck-mounted cranes, no loads shall be lifted over the front area except as approved by the crane manufacturer."

ANSI Requirements

ANSI also prohibits lifting over the cab of a truck-mounted crane except when approved by the manufacturer.

ISO 9000 Quality Management

An essential element is the development of a customer feedback system of pertinent hazard information on a continuous basis. This would include chronic loosening of bolts that lead to a failure mode. With such traceability at hand, manufacturers could then alert other users as to the problem *before* other failures occur.

Other References

See manufacturer's maintenance manual for the crane being inspected.

Military Specifications and Standards

1. *MIL-HDBK-60,* "Threaded Fasteners—Tightening to Proper Tension," 1990.
2. *MIL-N-25027G,* "Nut, Self Locking" (Service Temperatures 250 to 800 degrees F.), May 1995.
3. *MIL-F-18240E,* "Fastener Element, Self-Locking, Threaded Fastener," 250 degrees F. Maximum, December 1989.

Suggested Design Criteria

A safety factor sufficient to prevent unintentional over-stressing from overloading at any compass point on the turntable assembly should be included in design.

Maintenance manuals should require that scheduled routine checks of the turntable assembly be made. The manual should set forth the required torque for connecting bolts so they can be checked to maintain structural integrity.

The inclusion of a load-moment indicator (LMI) as standard equipment would avoid any overloading that might over-stress the turntable assembly.

Representative Litigation

Plaintiff Renkin; occurred April, 1981; Superior Court, Snohomish County, Washington, #83-2-02744-4.

> Injured sustained injury causing paraplegia when struck by crane boom that disconnected from pedestal when outriggers began to sink into ground as the partially extended boom was lowered to unload concrete pipe. Alleged that turntable was not strong enough for foreseeable loadings, causing the bolts securing the pedestal to break. Case settled.

Plaintiff Stephens; occurred November 17, 1977; State Court, Kent County, Michigan, #80-31996-CK.

> Injured's foot crushed when lifting punch press with mislabeled weight, causing crane to upset. Boom fell when the turntable raceway bolts broke. Case settled.

18

Tower Cranes

Upset Of Self-Raising Tower Cranes

Definition

The most hazardous operating cycle of a self-raising or climbing tower crane is when it is being raised or lowered from one level to another.

Description

When a new section of tower needs to be added to raise the upper horizontal boom, counterweight, and cab unit on some types of self-raising tower cranes, the upper unit is unbolted from the vertical tower and the new section inserted. During this procedure, if the temporary bolting or latching mechanism to hold the upper unit in place is not properly secured, the horizontal upper unit can become unbalanced and fall as the lifting trolley on the boom moves the new tower section into place. (See Figure 3-11.)

The substitution of lower strength connecting bolts can cause failure under stress. Additionally, a failure in the turntable assembly can also cause undocking of this type of crane. (See Chapter 17, Turntable Undocking.)

Risks Presented by Hazard

If this type of crane topples from the top of a skyscraper under construction, both the operator and all the people on the street below are in jeopardy.

Available Hazard Prevention Measures

A written standard operating procedure and a checklist for the climbing of this type of crane is necessary to prevent this hazard. The crane operator, riggers, and the person supervising the raising must each have a copy of the checklist so they can simultaneously follow the procedure step-by-step.

Communication either by telephone or radio should be available to verify completion of each step of the checklist before the next step is taken.

Connecting bolts and pins must be examined prior to erection to confirm that they meet the manufacturer's standards and are not dangerous substitutes.

Raising or lowering operations are best conducted on weekends or during other daylight periods when there is minimum exposure to the public in adjacent areas.

OSHA Requirements

OSHA has no specific requirements on this hazard, other than OSHA Subpart C, General Safety and Health Provisions, 1926.20(a) and (b) requirements for a safe workplace.

ANSI Requirements

ANSI has no specific guidelines on climbing procedures.

Other References

1. *Crane Handbook* by D. E. Dickie of Canadian Construction Safety Association of Ontario (1975) discusses erection, climbing, and dismantling in Chapter 7.

2. Appropriate manufacturers' operator's manuals should be reviewed to develop written step-by-step standard operating procedures for use by the foreman, riggers, and crane operator in raising operations.

3. *Safety and Health Requirements Manual,* EM 385-1-1 (U.S. Army Corps of Engineers, November 3, 2003), Section 16.E.02. requires: "Cranes shall be erected in accordance with the crane manufacturer's recommendations and the applicable ANSI/ASME standard.

   a. The manufacturer's written erection instructions and a list of the weights of each component to be erected shall be kept at the site.

   b. Erection shall be performed under the supervision of a qualified person.

   c. An AHA shall be developed and implemented for the erection procedures. The analysis will include a plan that shows:

      (1) The location of the crane and adjacent buildings or towers, overhead power and communication lines, underground utilities;

      (2) Foundation design and construction requirements; and

      (3) When the tower is erected within a structure, the plan shall show clearances between the tower and the structure and bracing and wedging requirements.

   d. Wind velocity at the site at the time of erection shall be a consideration and may be a limiting factor that could require suspending the erection operation.

   e. Before crane components are erected, they shall be visually inspected for damage. Damaged members shall not be erected."

Suggested Design Criteria

The design process should include a system-safety, fault-tree hazard analysis at time of design and a field service test to assure for comprehensive safety evaluation prior to the sale of each model to eliminate design defects.

Operating manuals should include step-by-step raising and lowering procedures that specifically list who does what for each step and a communication system to verify that each step has been completed before the next is undertaken.

Representative Litigation

Plaintiff Dearing; occurred April 20, 1985; Superior Court, King County, Washington, #86-2-00580-3.

Injured sustained psychological problems when the crane was being improperly raised to higher level. The crane operator was unaware that the raising crew had removed all the bolts from the turntable and that the horizontal boom, counterweight, and cab unit was precariously balanced on top of the vertical crane tower. When the operator lifted the section to be inserted in the tower, the unbalanced crane upset backwards and fell 400 feet. Miraculously, the injured sustained only minor physical injuries as the collapse of the downward-pointing boom absorbed most of the energy. Case settled.

Footing Failure on Tower Cranes

Definition

A tower crane can topple if placed on a poor foundation.

Description

A very important component in the operation of a tower crane is the footing upon which it stands. Unstable soil can render a crane inoperable very quickly. If a tower crane relies upon the structure being build for support, caution must be taken to assure that if concrete is used, that it has cured and attained its desired structural strength.

Risks Presented by Hazard

Serious injury or death can occur with little warning to those in the path of the fall of a tower crane.

Available Hazard Prevention Measures

A soils engineer should be responsible for designing and approving footings for tower cranes, and a structural engineer should be responsible for assuring that the structure supporting the tower crane is adequate for the loads that will be imposed.

OSHA Requirements

OSHA has no specific requirements on this hazard, but OSHA Subpart C, "General Safety and Health Provisions," 1926.20(a) and (b) require a safe workplace.

ANSI Requirements

ANSI B30.3, *Hammerhead Tower Cranes,* Section 3-1.1.1.c states that all load-bearing foundations shall be constructed in accordance with the manufacturer's recommendations.

Other References

1. *Crane Handbook* by D. E. Dickie of Canadian Construction Safety Association discusses footing failure in Chapter 7, (October, 1975).

2. Soils and civil engineering handbooks will provide additional guidance on various soil strengths.

3. *Safety and Health Requirements Manual,* EM 385-1-1 (U.S. Army Corps of Engineers, November 3, 2003), Section 16.E.01 states: "All load bearing foundations, supports, and rail tracks shall be constructed or installed in accordance with the crane manufacturer's recommendations and the applicable ANSI/ASME standard."

Suggested Design Criteria

A registered professional civil engineer who is a specialist in soils and foundations should design, sign drawings, and oversee placement of footings for a tower crane.

A registered professional structural engineer should review the drawings to determine if the structure supporting the crane is adequate. When reinforced concrete is used, the structural engineer should have sole responsibility in determining when the new structure can safely support the crane.

19

Overhead Bridge Cranes

Exposed Electrical Trolleys on Overhead Bridge Cranes

Definition
Energized electrical conductors are often exposed and accessible on overhead bridge cranes.

Description
Contact with energized electrical conductors generally occurs when painting, repairs, or other maintenance is being done, particularly when repair personnel attempt to cross from one overhead crane to another on the same rails or adjacent rails.

Risks Presented by Hazard
The risk to maintenance personnel is great, and severe electrical shock and/or a serious fall usually result.

Available Hazard Prevention Measures
Guard all exposed and accessible energized electrical conductors. Completely enclosed trolley systems are being marketed that effectively cover bare, uninsulated, energized electrical conductors.

OSHA Requirements
OSHA 1910.179(g)(1) and (2) and Marine Terminal Standard 1917.45(f)(9) require that conductors be enclosed so as to prevent contact under normal operating conditions and to comply with OSHA 1910, Subpart S.

ANSI Requirements
ANSI B30.2.0, *Overhead and Gantry Cranes*, Section 2-1.10.6, "Runway Conductors," states: "Conductors of the open type, mounted on the crane runway beams or overhead, shall be so located, or so guarded, that persons normally could not come into contact with them."

Other References
The National Electrical Code (National Fire Protection Association), Chapter 6, Special Equipment, Article 610, Cranes and Hoists, Paragraph 610.21(a), states: "*Locating or Guarding Contact Conductors.* Runway contact conductors shall be guarded and bridge contact conductors shall be located or guarded in a manner that persons cannot inadvertently touch energized current-carrying parts." (*The National Electrical Code* is also an ANSI standard.)

Suggested Design Criteria
Design and install enclosed trolley conductors so personnel whose tasks include repair, painting, and other maintenance cannot contact bare electrical conductors.

Representative Litigation

Plaintiff Sharpe; occurred August 12, 1974; Maricopa County, Arizona, #C315922.

> Glass installer was electrically burned and fell when walking on bridge crane rail enroute from one bridge crane to another. Alleged unsafe access route and unguarded live electrical conductors. Case settled.

Absence of Lockout Systems on Overhead Bridge Cranes

Definition

Overhead bridge cranes sometimes do not have lockout systems to prevent operation while maintenance or service is being performed.

Description

On overhead bridge cranes, the operator's cab is sometimes suspended from the underside of the bridge. From this position, the crane operator cannot see the top of the bridge where the trolley hoist moves back and forth. Painting of the ceiling of the building, changing of lights in the ceiling, making routine checks on the trolley hoist, or other maintenance or service activities are sometimes done from the top of the bridge or trolley hoist. The operator may be unaware of activities conducted in areas that cannot be seen from the cab.

Maintenance personnel may also be working along either of the rails on which the crane travels and may not be seen by the operator. Sometimes the rails are located within one or two inches of a support pillar for the building, and this narrow clearance creates a very dangerous pinch point when the crane travels down the rails past the pillar.

Generally, the switch boxes for the electrical systems that power these cranes are remotely located on the ground floor and immobilize the entire crane. There are no separate lockout switches for the hoist motor, trolley-travel motor, bridge-travel motor, or other components.

Risks Presented by Hazard

Maintenance and service personnel working on top of the bridge out of view of the operator may be crushed by the moving trolley or caught in the hoist if the trolley system is not immobilized.

Severe crushing can occur if someone is caught in the narrow clearance between the moving bridge crane and a building pillar.

Available Hazard Prevention Measures

To prevent the use of the trolley hoist while personnel are on the top of the bridge crane, a separate lockout switch should be provided adjacent to the operator's position so personnel going onto the top of the bridge can put their own personal lock on the switch to alert the operator that someone is on top of the bridge. Individual locks are necessary because the crane operator has no way of knowing who may or may not be on or off the crane.

A similar arrangement is needed to immobilize crane travel. Temporary rail stops can be placed on the tracks to prevent crane travel in areas where other work is being performed. Conspicuously-placed flags should be used to advise the operator that travel in that area is restricted by rail stops.

Include a written lockout procedure in the operator's manual and in maintenance procedures.

OSHA Requirements

OSHA 1910.147 sets forth requirements for lockout devices. OSHA 1910.179(l)(2) discusses maintenance requirements.

ANSI Requirements

ANSI B30.2.0, *Overhead and Gantry Cranes*, Section 2-2.3.2, requires both protective rail stops and a lockout procedure.

Other References

Safety and Health Requirements Manual, EM 385-1-1 (U.S. Army Corps of Engineers, November 3, 2003), Section 12.A.07, Hazardous Energy Control Plan, states:

> a. Hazardous energy control procedures shall be developed in a hazardous energy control plan.
>
> b. The plan shall clearly and specifically outline the scope, purpose,

authorization, rules, and techniques to be utilized for the control of hazardous energy, including, but not limited to, the following:

(1) a statement of the intended use of the procedure;

(2) means of coordinating and communicating hazardous energy control activities;

(3) procedural steps and responsibilities for shutting down, isolating, blocking, and securing systems to control hazardous energy;

(4) procedural steps and responsibilities for the placement, removal, and transfer of lockout and tagout devices;

(5) procedural steps and responsibilities for placing and tagging, and moving or removing and untagging, protective grounds;

(6) requirements for testing the system to verify the effectiveness of isolation and lockout and tagout devices;

(7) a description of any emergencies which may occur during system lockout or tagout and procedures for safely responding to those emergencies;

(8) requirements when authority for removal of hazardous energy control devices must be transferred from the authorized employee to another individual, and the names of the individuals qualified for receiving such transfer; and

(9) the means to enforce compliance with the procedures.

Suggested Design Criteria

Locate the master electrical switch that turns the entire crane off at the ground-floor stairs leading to the crane.

Adjacent to the operator's cab, locate a lockout switch that will accommodate multiple locks for each component — the hoist motor, the trolley-travel motor, the bridge-travel motor, etc.

Design the rails to accommodate temporary rail stops to block crane travel. A hole in the web of the rail placed every ten or fifteen feet along the length of the rail would satisfy this need.

Provide a written lockout procedure in the operator's and maintenance manuals.

Representative Litigation

Independent maintenance contractors or their employees who have been crushed by moving cranes have filed claims against crane owners, and those employed by crane owners have filed claims against crane manufacturers and/or installers alleging the crane was inherently dangerous without lockout systems.

Plaintiff Nicholas; occurred June, 1981; Eastern District, Northern Division, Michigan, #84-CV-9260-BC.

> Deceased electrician crushed between vertical roof support post and passing overhead crane while working on elevated rails. Alleged that manufacturer failed to include lockout system. Case settled.

Plaintiff Mudlo; occurred March 22, 1974; Court of Common Pleas, Allegheny County, Pennsylvania, #76-04962.

> Welder injured by moving crane while out of the view of the crane operator. Alleged that operator's vision was restricted, no travel warning alarm, and no lockout system. Case settled.

Plaintiff Wolfe, occurred December 16, 1987; Circuit Court, Kent County, Michigan, #89-61587-NO.

> The injured, an employee of a contractor who was remodeling an operating plant, was standing on overhead tracks while welding and was struck by bridge crane moving equipment being operated by a plant employee from a floor-level suspended control box. Alleged that no pre-construction planning was done by the plant with the contractor to assure that contractor's employees would be protected by an effective lockout system during plant operation. Case settled.

Monorails and Underhung Crane Stops

Definition

Monorail systems sometimes have inadequate stops, and the monorail's extensions can become disconnected and fall.

Description

A monorail crane consists of a fixed track suspended from the ceiling. Sometimes another traveling monorail track with a suspended traveling hoist is hung from the monorail, giving greater flexibility to either extend the hoist to each side or beyond the end of the fixed monorail. (See Figure 3-8.) When loads are being moved to the outer limits on either of these rails, the stops, which receive severe pounding from the powered travel mechanism and the load, sometimes fail.

Risks Presented by Hazard

Workers in the immediate vicinity of a monorail can be crushed by either a falling traveling monorail track or by its suspended traveling hoist and its load if stops are inadequately secured and fail.

Available Hazard Prevention Measures

Use well-designed rail clamps with good bumpers and good anchoring systems to prevent the suspended traveling monorail or its hoist from going off the end of either track.

OSHA Requirements

OSHA 1910.179(j) requires periodic inspection that includes examination of stops and worn bolts, etc. but does not specifically refer to monorails.

ANSI Requirements

ANSI B30.11, *Monorail Systems and Underhung Cranes*, Chapter 11-1, "Construction and Installation," contains significant discussions on track support.

Other References

Many machine design textbooks have design principles for developing safe rail stops.

Suggested Design Criteria

Design of monorail stops needs to include a pin through the web of the monorail and a clamp and shock absorber so they will not loosen from repeated impact.

Representative Litigation

Plaintiff Wood; occurred August 30, 1985; Circuit Court, Oakland County, Michigan, #86-319-384-N.P.

> Deceased crushed by third-tier assembly and hoist when it fell from second-tier assembly from which it was suspended due to an inadequate one-bolt stop. Alleged that two-bolt stop was required. Case settled.

21

Guyline Anchor Systems

Data for this chapter was provided by A. J. Scardino of Sigma Associates, Ltd., 105 Timber Ridge Boulevard, Pass Christian, Mississippi 39571, (601)452-4866.

Definition

Guylines and anchors are often used to support A-frames, gin poles, derricks, and other lifting systems.

Description

Mast-type devices held in place with guylines and anchors experience constant and varying dynamic loading conditions. For example, an oil well derrick (work-over rig) has an elevated work platform with people working on it (Figure 21-1) and pipes are often stacked inside the derrick tower; and A-frames, gin poles, and other derricks must withstand jarring, wind loading and other various forces.

Investigations of fatal accidents involving the stability of oil well derricks by the U.S. Department of Labor show that component failures of either guylines and/or anchors were the cause of oil well derricks overturning. Oil well derricks are a very sophisticated application of raised-mast lifting systems that have evolved from A-frames, gin poles, and derricks in general.

Risks Presented by Hazard

If the guyline breaks or the anchor pulls from the ground, the supporting mast can fall, causing injury to those on the work platform or on the ground below.

Available Hazard Preventive Measures

Lessons learned from stability systems in the oil industry need to be applied to a broad range of raised-mast lifting devices that present unreasonable dangers when they lose stability and topple because of a failure of the guylines and/or anchors.

The stability of oil well derricks, gin poles, and A-frames are dependent upon the foundation, anchors, and guylines.

1. *Foundation.*

   a. The area should be graded, leveled, and maintained so that oil, water, drilling fluid, and other fluids will drain away from the working area.

   b. Safe bearing capacity shall be determined from the use of an appropriate table, soil core test, penetrometer test, flat plate test, or other suitable soil test. When surface conditions are used to determine bearing capacity, care must be exercised to insure that the soil is homogeneous to a depth at least twice the width of supplemental footing used to support the concentrated load.

   c. Supplemental footing shall be provided to distribute the concentrated loads from the mast and support points of the oil well derrick, gin pole, or A-frame. The manufacturer's load distribution diagram will indicate these locations. In the absence of a manufacturer's diagram, the supplemental footing shall be designed to carry the maximum anticipated hook load, the gross weight of the mast, the mast mount, the traveling equipment, and the vertical component of guywire tension during operational loading conditions. These footings must also support the mast and mast weight during mast erection.

d. Wellhead cellars present special foundation considerations. In addition to obvious problems with water and fluid seepage, cellars also require unique mast support considerations. These should be analyzed by a qualified person to insure that an adequate mast foundation is provided.

e. Small settlements (soil subsidence) at the beginning of rig-up are considered normal. External guywires should never be used for plumbing the mast. Rig foundations, guywire anchors, and guywire tension should be checked at each tower (shift) change.

2. *Anchors.* There are both manufactured and in-house, shop-made anchors.

  a. Manufactured anchors.

   1) There are four basic types of manufactured anchors. The screw or helix anchor, expanding plate anchor, flat plate anchor, and the pivoting anchor. Holding capacity of these anchors varies. Detailed information specifying anchor holding capacity, comparison charts with illustrations, and characteristics of each design may be found in Section 2 of *Guidelines on the Stability of Well Servicing Derricks* by A. J. Scardino. (See Other References at the end of this chapter.)

   2) When installed in conformance with manufacturer's specifications, evidence thereof provided, the requirement for individual pull testing has been satisfied. *Caution: It should continually be emphasized that the anchor is only one component of the Stability System.*

   3) Screw (helix) type anchors have a direct correlation between anchor capacity and the torque required to install the anchor. Following the manufacturer's specific torque recommendations, with proof thereof, is a valid method of determining anchor holding capacity. Torquing, according to manufacturer's specifications, is an acceptable, non-pull test method of determining anchor capacity.

  b. Shop-made (in-house fabricated) anchors.

   1) These anchors should be designed by a registered professional engineer conforming to accepted engineering practices. Written procedures shall be established for installation.

   2) These shop-made anchors should be proof-tested for structural integrity and holding capacity. Records shall be maintained as to testing protocols and holding capacity based on soil type.

   3) Individual pull testing will not be required if anchors are installed in accordance with written procedures. Proof thereof will be required of installation protocols and proof-tested holding capacities.

3. *Guywires.*

  a. The manufacturer's guyline recommendations shall be followed. In the absence of manufacturer's recommendations, the following location diagram, Figure 21-2, may be used.

  b. Each zone requires an anchor of different holding capacity. (See Figure 21-3.) If anchors are located in more than one zone, then all anchors should be of the capacity required for the greater capacity zone; i.e., if one anchor is located in Zone C and the remaining anchors are related in Zone D, all anchors shall meet the holding capacity specified in the chart for Zone C.

c. All guywires, as indicated by the manufacturer's diagram, should be in position and properly tensioned prior to commencing any work.

d. In the absence of the manufacturer's recommendations or where manufacturer's recommendations for masts cannot be implemented, the diagram in Figure 21-4 may be used.

e. Other guying patterns may be used. They must, however, be based upon sound engineering principles, determined by a qualified person. These recommendations should be posted on the mast, in a weather proof container, and should state the loading conditions for which they were prepared.

f. Guywires should be 6 x 19 or 6 x 37 class, regular lay, made of improved plow steel (IPS) or better with independent wire rope core (IWRC), not previously used for any other application. Double saddle clips should be used. Wire rope should be installed in accordance with the manufacturer's recommendations.

4. *Visual Observations.* There are characteristic visual observations that can be accomplished which serve as indicators of the stability of an oil well derrick, gin pole, or A-frame. They include but are not limited to the following:

a. The foundation supports the rig, substructure, and all applied loads, while in an operational mode, without excessive movement, basically in a level and plumb configuration.

b. No large movement is observable between the mast support structure and the rotary/setback support structure when the slips are set and the load is removed from the mast, or vice versa.

c. The empty travel block hangs plumb with the centerline of the wellbore and the mast support structure remains level.

d. The mast support structure and/or substructure does not lean to one side more than the other when the

load is applied. The guywire on one side should not be noticeably taut while the guywire on the opposite side is slack.

e. The guywire anchor(s) shows no visible signs of movement during the loading and unloading of the system while in its operational mode.

The chart presented in Figure 21-5 may be used as a guide to the pre-tensioning of guywires. This method is commonly referred to as the "Catenary Method" (guywire sag method).

OSHA Requirements

OSHA has not commented on specific requirements that address the stability of oil well derricks and servicing rigs, as well as A-frames, gin poles, and derricks in general.

ANSI Requirements

Derricks, ANSI B30.6-1969, doesn't comment on terms of specific requirements concerning stability systems.

Other References

1. *Specification 4E: Specification for Drilling and Well Servicing Structures* (American Petroleum Institute, Washington, D.C., 1988).

2. *Recommended Safe Procedures and Guidelines for Oil and Gas Well Servicing* (Association of Oilwell Servicing Contractors, Dallas, Texas, 1988).

3. *Accident Prevention Manual* (International Association of Drilling Contractors, Houston, Texas, 1990).

4. *Drilling Manual* (International Association of Drilling Contractors, Houston, Texas, 1979).

5. *Guidelines on the Stability of Well Servicing Derricks,* A. J. Scardino (Sigma Associates, Ltd., Pass Christian, Mississippi, 1990).

Suggested Design Criteria

Registered professional engineers should design cable-supported mast lifting systems, and they should sign and seal the drawings.

Figure 21-1 Oil Well Servicing Derrick

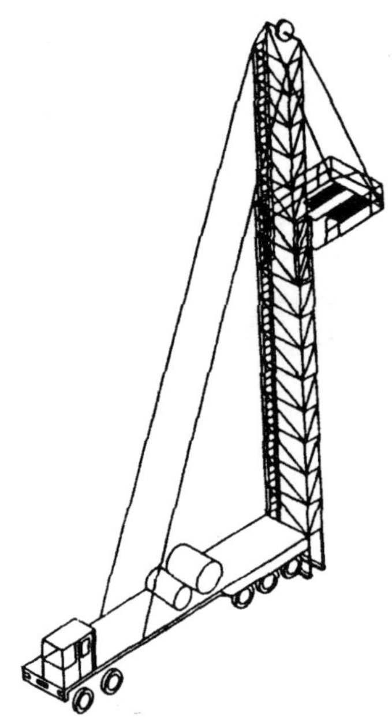

Figure 21-2 Recommended Anchor Locations

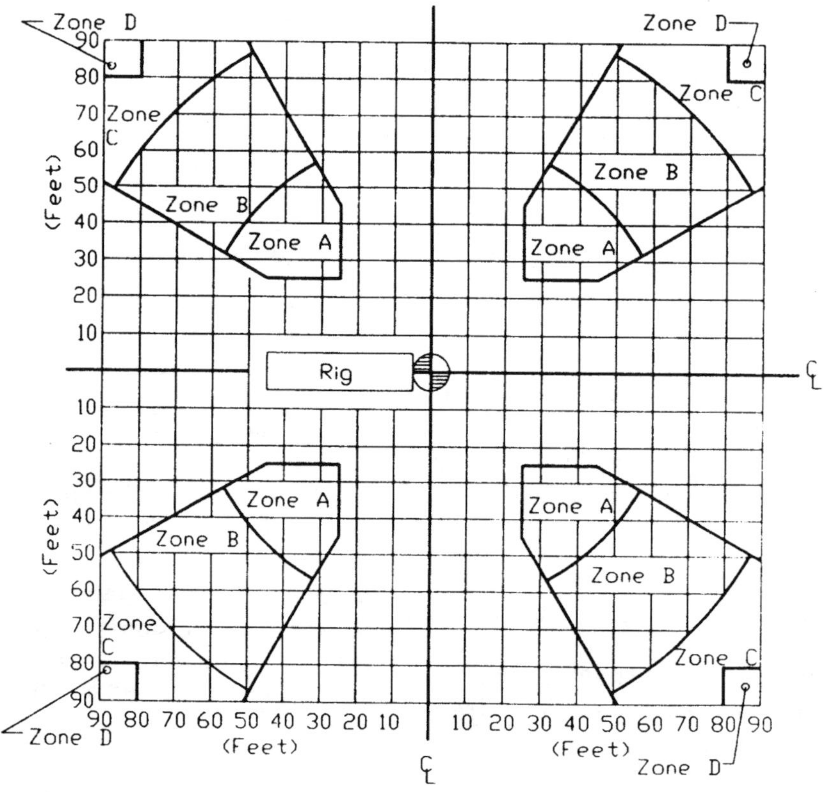

Figure 21-3 **Anchor Capacity Requirements For Each Zone**

ANCHOR CAPACITY (TONS)

Zone	Doubles Mast	Singles Mast	Pole Mast
A	15.6	7.0	7.0
B	11.5	5.0	5.0
C	9.0	5.0	5.0
D	7.4	5.0	5.0

Anchor capacities shown assume the following:

1. Adequate foundation support for mast and carrier.

2. Adequate crown-to-carrier internal load guys.

3. Maximum wind load – 70 MPH

4. Maximum hook load – as described elsewhere, this chapter.

5. Full rod and tubing setback (N/A for Pole unit).

Figure 21-4 **Anchor Location Diagram**

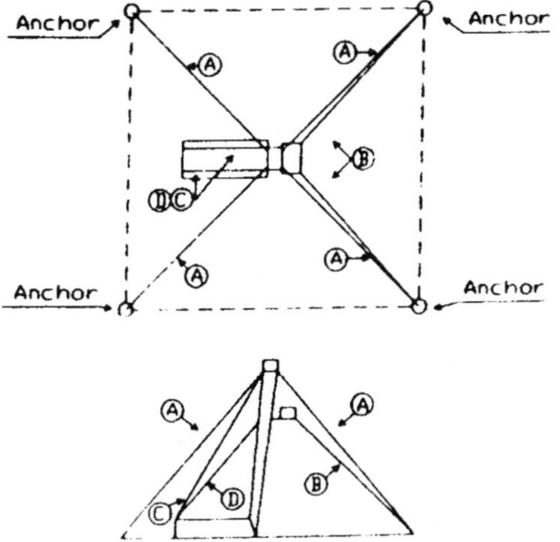

Figure 21-5 **Catenary Method**

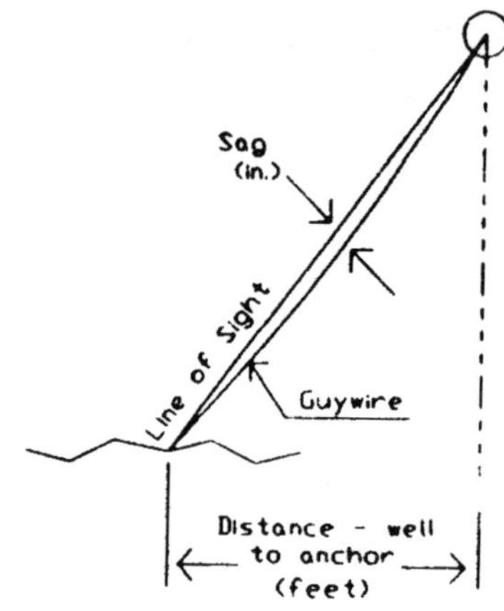

	GUYWIRE SAG (inches)					
	Pole Mast		Singles Mast		Doubles Mast	
Distance Well to Anchor (ft)	Tubing Board Guy	Crown-Ground Guy	Tubing Board Guy	Crown-Ground Guy	Tubing Board Guy	Crown-Ground Guy
40	-	4	4	4	6	5
60	-	6	8	6	12	8
80	-	10	15	10	17	11
100	-	14	22	14	26	15
120	-	18	32	18	32	21
Pre-tension (Pounds)	500	1000	500	1000	500	1000

22

Wind and Weather

Definition

A long raised boom is very susceptible to being blown over by high winds.

Description

Winds can topple a raised boom, and some tower cranes are vulnerable to this hazard.

Risks Presented by Hazard

A falling boom or tower endangers those in the immediate vicinity below.

Available Hazard Prevention Measures

All craning activities should be immediately discontinued when windy conditions arise.

Mobile cranes with long booms should be lowered to the ground when high winds are predicted.

Wind indicators are available that give warning of dangerous winds.

Manufacturers' instructions on preparation of tower cranes for high winds must be followed.

To reduce this hazard on tower cranes, the turntable should be allowed to rotate freely and weathervane with the wind.

OSHA Requirements

OSHA 1910.179(b)(4) sets forth requirements for wind indicators for overhead and gantry cranes.

ANSI Requirements

ANSI B30.3, *Hammerhead Tower Cranes*, Section 3-1.2.16, requires a wind velocity device.

Other References

1. *Crane Handbook* by D. E. Dickie of Canadian Construction Safety Association of Ontario (1975) discusses this more fully in Chapter 5, p. 193.

2. *Safety and Health Requirements Manual*, EM 385-1-1 (U.S. Army Corps of Engineers, November 3, 2003), Section 16.C.19.a. states: "Cranes/derricks shall not be operated when wind speeds at the site attain the maximum wind velocity recommendations of the manufacturer. Projects shall have adequate means for monitoring local weather conditions including a wind indicating device."

Suggested Design Criteria

The wind loading that the crane is designed to withstand should be clearly stated in the operator's manual.

The operator's manual for a tower crane should provide specific guidance on how to prepare the crane for wind storms, hurricanes, tornadoes, icing or other emergency weather conditions.

23

Maintenance and Renovation

The larger the crane, the longer its useful life expectancy. Small mobile cranes can remain in use for thirty or forty years. Large gantry and bridge cranes are sometimes still in use at certain facilities from ninety to over a hundred years. Considering this long life, the owner/user should keep an ongoing maintenance and renovation record, and the manufacturer should maintain an ongoing record of renovations and improvements to assure that the crane is updated with parts and components to keep it current with available safety technology.

Crane manufacturers and rental firms need to be vigilant in their review of purchase orders for replacement parts and/or components. Numerous orders to replace the same part or component may be a forewarning that the design is faulty and that failures will probably continue to occur that in the long run might cause injury or death. For example, frequent replacement of jib booms might indicate that a hazard exists that needs correcting. (See Chapter 14, Boom Failures and Disassembly.) It is not enough simply to fill the orders. The manufacturer should also determine why critical components are continually needing replacement.

The crane model, serial number, and original maintenance/parts manual are essential information, and all record-keeping systems must be maintained for the entire life of the crane. The steel, plastic, and other materials used in older cranes have been vastly improved through the years. Electrical and electronic control systems have evolved into new state-of-the-art technology almost overnight. Cranes and other functional production equipment lend themselves to modernization. Many older cranes have sound basic structure. When compared to the cost of a new crane, an old crane can be modernized for a modest investment, improving performance and reducing operating costs. In some instances a crane can be converted to a higher-rated capacity. It is mandatory to seek the manufacturer's assistance, as all conversions relating to the lifting capacity of the crane need to be consistent with both OSHA and ANSI requirements. For example, when a bridge crane is converted from a cab-operated crane to a pendant-operated crane to reduce labor costs, the following should be considered:

1. State-of-the-art suspended pendant control that includes:

   a. A festooned line that allows the pendant to be moved the full length of the bridge. This gives much more maneuverability to the operator to observe the lift at all times and stay clear of the lift while making sure that the travel of the crane endangers no one.

   b. Control buttons that are raised to indicate by touch the function of specific controls.

   c. An emergency stop switch that shuts down the entire crane and renders the crane electrically inert, with circuitry that requires resetting the main power circuit breaker to the crane before crane operations can be resumed.

2. Reliable limit switch system to prevent over-winding.

3. Redundant, audible alarm to warn the operator that safe vertical travel limit has been reached.

4. Travel alarm to warn those working adjacent to the crane or the load of impending movement. Automatic delays are available that momentarily sound the alarm before movement to give people time to move to a safe location. (See Chapter 11, Vision.)

5. Lockout system to prevent travel or movement of the crane when maintenance is being done.

New cranes are often more reliable than older ones, and manufacturers have developed improved parts and components that could be utilized on some older models of the same make. Owners of older cranes who would benefit by updating their cranes should be notified of available new parts and components by the manufacturer and the maintenance/parts manual should be updated. By reviewing the maintenance and renovation records and assuring that the maintenance/parts manuals are current, the manufacturer's sales or service representatives can play a critical role in assuring that the crane is maintained and has been updated in a safe manner and that repairs and renovations were made with proper parts. Programs of this nature are cost-effective for both the manufacturer and the crane owner as renovation and updating provides an ongoing market for the manufacturer and a more reliable and productive piece of equipment for the user. Since the advent of load-measuring indicators, particularly those that limit lifts to a safe load, the risk of upset has been dramatically reduced. Many crane improvements are adaptable to a retrofit program. Retrofitting not only enhances the safety of the crane, but provides a profitable market for crane manufacturers. Crane manufacturers' sales representatives should also be sure that distributors have an effective system of assuring that customer service systems are able to verify that parts

and/or components are properly identified so incompatible parts and/or components will not be installed. For example, failure to use the steel pins provided by the manufacturer for connecting latticework-boom sections could result in boom collapse. A simple substitution many times appears to be quite harmless, but it can create a hazard with catastrophic consequences.

When a crane-related injury occurs, a thorough investigation should be initiated by the local distributor. The manufacturer should be notified. Under the Freedom of Information Act, copies of OSHA or MSHA injury reports can be obtained from the U.S. Government for only a copying charge. The local distributor should obtain this information and forward it to the manufacturer. When a particular part or component fails and may have been the cause of the occurrence, it should be retained in a safe place for future examination. Disposal of damaged parts or components may make it difficult to make a valid hazard analysis or injury investigation. After a competent analysis has been made to identify the hazardous circumstance and necessary hazard prevention measures, a summary should be made by the manufacturer to inform its designers, distributors, and all crane owners/users of the hazardous circumstances, including specific recommendations for necessary hazard prevention measures.

A most important item in safe crane maintenance and renovation is to keep crane owners and operators informed of the injury experience of their cranes. Failure to publicize hazard information allows armed hazards to remain undetected. The two greatest obstacles in overcoming hazards are:

1. The temptation to blame the victim and/or the operator and the failure to identify any hazards that may have been the real culprits.

2. The temptation to keep hazard information a secret to avoid having to make changes on similar equipment.

Management should take prompt remedial action when a hazard is first identified. When a hazard is identified after fifty cranes are sold, it is best to fix the fifty cranes immediately and change the design of future models than it is to wait until five hundred defective cranes are in the marketplace and hope to cover any losses with insurance. The following little verse summarizes why insurance is the most costly method of controlling hazards:

> It is unwise to keep a hazard, but it is worse to pay for the harm it causes.
> When you pay to remove the hazard, it costs you a little money — but that is all.
> When you do not pay to remove a hazard, you sometimes lose everything, because the hazard was capable of causing harm.
> If you keep the hazard, it is well to add something for the risk you run.
> And if you do that, you will have more than enough to pay for the removal of the hazard.

Crane safety is dependent upon good maintenance and renovation, which includes informing crane owners and operators about current hazard information. Safety professionals and managers whose operations include the use of cranes can also play a very important role by seeing that cranes are maintained in a safe manner by:

1. Assuring that each crane has a thorough annual load testing and inspection by someone recognized by the U.S. Department of Labor. (See OSHA 1926.550(a)(6).) When a crane fails during inspection, it must not be operated until necessary repairs are made.

2. Auditing maintenance records to assure that they are consistent with the manufacturer's recommendations.

3. Assuring that the manufacturer is informed in writing of any renovations or remodeling that is being contemplated and securing approval before work is commenced.

4. Reviewing crane literature and crane sales brochures, investigating available after-market operator aids, and attending crane trade shows to evaluate what updates would be advantageous for safer and more reliable crane use and greater productivity.

5. Researching injury reports, litigation, and other crane hazard data that is applicable to the type of cranes being used on their operations to learn by the experience of other users.

6. Investigating each crane-related injury and preparing a written hazard analysis. In the event of a serious or fatal occurrence, it is prudent to retain a competent safety engineer who is knowledgeable in crane hazards to do an independent hazard analysis that includes:

   a. Description of occurrence.
   b. Description of accident site.
   c. Statements of personnel interviewed.
   d. Documents, references, statements, photographs, and other data relied upon.
   e. Discussion of the hazard and its history, including previous occurrences and the circumstances that activated the hazard.
   f. Available hazard prevention measures, including pertinent design improvements and operating procedures.
   g. Recommendations as to how appropriate hazard prevention measure are to be instituted.

7. Providing continuing training for crane operators, riggers, users, and those supervising crane operations that addresses the limitations of crane use and current or revised operating/maintenance practices to assure for safe crane use, maintenance, and renovation.

24

Personnel Requirements

Competency of Operators

As cranes become more sophisticated and are able to lift heavier loads, higher, farther, and faster, more and more electronic systems to monitor every aspect of operator performance are being included in design. The day of total reliance upon "seat-of-the-pants" operator skills is gone. Human response is not fast enough to cope with many of the rapidly changing circumstances involved in crane operation. Today's crane can be compared to an airplane, not just in terms of cost, but in its complexity of operation. They require well-qualified professional operators. Some employers have begun to do in-house licensing of crane operators based upon the following:

1. Education level.
2. Apprenticeship training and work experience.
3. Classroom training on crane safety.
4. Knowledge of crane safety requirements and references.
5. Physical qualifications:
   a. Age (mature and intelligent).
   b. Vision. ANSI/ASME B30.5, *Mobile and Locomotive Cranes*, Section 5-3.1.2(b)(1) requires: "Vision of at least 20/30 Snellen in one eye and 20/50 in the other, with or without corrective lenses" and Section 5-3.1.2(b)(2) requires: "Ability to distinguish colors, regardless of position, if color differentiation is required for operation...."
   c. Hearing. ANSI/ASME.5, *Mobile and Locomotive Cranes*, Section 5-3.1.2(b)(3) requires: "Adequate hearing, with or without hearing aid, for the specific operation...."
   d. Physical stamina.
   e. Good coordination, reaction, and tested skill level.
   f. No history of heart problems or other ailments that could produce seizures.
   g. Emotional stability.
   h. Absence of drug and/or alcohol abuse and addictions, etc.
6. Record of injury or property damage while operating cranes.

There is a definite trend for certification of crane operators. The following now require certification:

1. Countries: Canada (Ontario), Germany, Holland, Norway, and Sweden.
2. States: California, Connecticut, Illinois, Massachusetts, Maryland, Montana, New Jersey, New York, and Oregon.
3. U.S. cities: Chicago, Los Angeles, New York City, and Washington, D.C.

4. U.S. agencies: U.S. Navy. (The U.S. Army Corps of Engineers has proficiency qualification requirements.)

The criteria for licensing crane operators ranges from an in-depth program that incorporates age requirements, specific training, skill tests, a written test, physical examination, personal profile, and an injury/damage-free experience to those that appear to have no real value except to raise revenue.

There is limited data available as to the results of requiring crane operator certification, but Ontario, Canada, has compiled some positive figures:

1. During the decade prior to formal training and licensing of crane operators, 1969 through 1978, Ontario had 8.7 construction fatalities per 100,000 workers per year, 3.59 of which were related to cranes.

2. From 1978 to the present Ontario has had 4.3 construction fatalities per 100,000 workers per year, 1.40 of which were related to cranes.

These statistics indicate that formal training and crane operator licensing are important factors in reducing crane-related deaths and injuries.

To assure for uniformity so that the qualifications and performance of crane operators are consistent nationwide from job to job, some type of nationwide crane-operator certification and licensing is needed. The Specialized Carriers and Riggers Association (SC&RA) and American National Standards Institute (ANSI B30 Committee) are currently drafting standards for crane operators.

It should be noted that "to license" and "to certify" mean two different things. To license is to grant permission to engage in a specific professional practice, with authority to enforce compliance with certain standards. To certify is to say that certain requirements or standards have been met. To be able to license and certify, those with authority must have standards that are meaningful. Several questions need to asked: "Who will be the governing authority: Federal, State, County, Municipal, Board? Are standards consistent throughout the nation? Any licensing or certification program will only be as effective as the standards that are set. If minimum consensus standards are adopted, little reduction in death or injury can be anticipated. Standards should include:

1. Description of the duties, responsibilities, and authority of a crane operator.

2. Types of cranes he may operate, listing:
   a. Capacity.
   b. Classification (hydraulic, lattice-work boom, bridge, etc.).
   c. Activity (hoisting, dragline, pile driving, etc.).
   d. Boom length.
   e. Type (crawler, rough-terrain, tower, etc.).

3. Skill test.

   The state of New York, in licensing crane operators, uses video tapes to verify an operator's judged ability to perform the State's required operating procedures.

4. Special training.

5. Level (master first class, second class, third class, apprentice, etc.).

6. Revocation and discipline.

7. Keeping a log of operating time (who, when, on what equipment).

8. Renewal.

9. Medical testing that requires a comprehensive assessment of physical and emotional characteristics.

10. Both a written and practical examination to determine actual knowledge of crane use and hazards. Determine whether operator has sufficient language skills. Special bilingual requirements may be necessary where minority groups speak a foreign language. In its *Safety and Health Requirements Manual*, EM 385-1-1 (U.S. Army Corps of Engineers, October, 1992) Appendix G reads:

    Crane operators shall pass a written examination which demonstrates their knowledge of the following:

    (1.) Responsibilities of operator, rigger, signal persons, and lift supervisor.

    (2.) Knowledge of USACE crane safety requirements and the crane's operator manual.

    (3.) Ability to determine the crane

configuration, compute the size and shape of loads, and determine the crane's capacity using the load chart.

(4.) Use and limitations of crane operator aids.

(5.) Inspection, testing, and maintenance requirements.

(6.) Determination of ground conditions and outrigger and matting requirements.

(7.) Crane set-up, assembly, dismantling, and demobilization procedures.

(8.) Requirements for clearance from power sources.

(9.) Signaling and communication procedures.

(10.) Factors which reduce rated capacity.

Crane operators shall pass a practical operating examination which demonstrates their ability to perform the following:

(1.) Inspecting the crane.

(2.) Establishing a stable foundation and leveling the crane.

(3.) Raising, lowering, extending, retracting and swinging the boom.

(4.) Raising and lowering the load line.

(5.) Attaching the load, holding the load, and moving the load.

(6.) Reading load, boom angle, and other indicator devices.

11. Qualifications of the examiners.

Training

Formal classroom training for crane operators and crane users is a very important factor in reducing crane-related injury and property damage and must be taught by competent instructors. Section 16.C.04.c. of *Safety and Health Requirements Manual*, EM 385-1-1 (U.S. Army Corps of Engineers, October, 1992), requires that its own hired employees who are

crane and derrick operators shall have a 24-hour course in operation and safety every three years. Comprehensive instructor's guides for those responsible for presentation need to include the following:

(1.) Prerequisite qualifications of prospective students to assure that they have the necessary background.

(2.) Bibliography of reference materials such as textbooks, literature, safety standards, and safety codes that need to be made available to the student.

(3.) Course outline for each hour of instruction to include basic hazard information and hazard prevention.

(4.) List of specific crane hazards. This may take several hours of classroom discussion.

(5.) Exercises on using crane load charts.

(6.) Use of load-measuring indicators and other crane operator aids.

(7.) List of student homework assignments.

(8.) Written course examination.

Competency of Riggers, Signalers and Others

Riggers, signalers, and others who work with cranes should have qualifications and training similar to those of the operator. Just as an unqualified operator can make a life-threatening error during lifting operations, the inappropriate actions of an inexperienced rigger, signaler, or anyone else involved in lifting operations can cause injury.

General Hazard Prevention Procedures

Top Management Involvement

The active participation of the chief executive officer and executive staff must be involved in crane safety. They are often completely naïve about the inherent crane hazards and that it is their responsibility to set priorities for the prevention of hazards. Management needs to ensure that all cranes to be purchased or leased are equipped with necessary safety accessories, such as load moment devices (LMIs) and anti-two-block devices. If it is anticipated that a crane is to be used next to powerlines, it should be equipped with proximity alarms and insulated links and a Boom Buoy, as the crane operator needs all of the help he or she can get as "thin-air" clearances are not enough. Further, in the selection of a crane it is prudent to ask the manufacturer about liability complaints based on the particular crane's hazards. Cranes being rented should be screened to see that they are properly equipped for their intended use and, most important, that the operator is qualified to operate that particular crane and holds a valid certification.

In recent crane disasters covered in our national media, the reported findings of investigations have often been critical of the construction manager's conduct. No longer can management of a project *just* ensure that a facility is being built to meet design specifications and that safety is the responsibility of the various contractors. *Oversight of safety is a non-delegable duty.* Invariably, serious crane accidents end up in litigation, which spares none of the various defendants their individual responsibility, based upon the facts of the occurrence. Slowly but surely, these court actions are increasing the emphasis on safe cranes and their safe use. This is becoming a top management priority, replacing the old belief that the worker has the responsibility to work safely with dangerous equipment.

Management that desires to learn more about the prevention of hazards involved with the use of cranes in specific circumstances can contact the Hazard Information Foundation, Inc. (HIFI). This is a not-for-profit public service organization dedicated to obtaining and maintaining accessible and comprehensive hazard information and available safety engineering alternatives to either eliminate or minimize hazards from causing injury, death, or property damage. For further information, see its web page, *http://www.hazardinfo.com.*

Pre-Construction Planning

Analysis shows that most crane-related injury and property damage could have been prevented had the safe use of cranes been considered at the pre-construction planning meeting, where hazards involving powerlines, lack of communication on blind lifts, lifting capacity, use of cranes and derricks on barges, special circumstances requiring two or more cranes to lift a single load, and other anticipated hazardous activities as outlined in this book should be topics of discussion.

Prior to the planning meeting, the safety professional should review the proposed project, find out the types of cranes that will most likely be used for specific activities, make an outline of specific hazards that may be encountered, and list the appropriate hazard prevention measures that should be discussed at the planning meeting, using this book as a guide.

Job Hazard Analysis

When cranes arrive at the job site, the work circumstances may not be the same as anticipated at the pre-construction planning meeting. Before actual crane operations are begun, an on-site job hazard analysis should be made to determine whether the crane is right for its intended use, whether the operator is qualified, and whether the supervision, riggers, signal persons, and others who will be involved are informed and knowledgeable as to the tasks to be performed.

Prior to making any lifts, crane operators, lift supervisors, riggers, and signal persons should meet together and review the following:

1. Exact size and weight of the loads to be lifted, as well as all crane and rigging components that would add to the weight.
2. Crane position, height of the lift, the load radius, and boom length and angle for the entire range of the lift.
3. Lift points, rigging procedures, and hardware requirements.
4. Ground conditions, outrigger or crawler track requirements, and adequacy of mats or cribbing to assure for stability.
5. Environmental conditions under which lift operations are to be stopped.
6. Coordination and communication requirements.

The job hazard analysis should also include a detailed examination of the crane as outlined in the Mobile Crane Hazard Analysis Worksheet as outlined in the appendix.

When inadequate planning has been done, this on-site job hazard analysis provides a necessary second-chance to assure that crane operations can be done safely. (See bibliographic listings 108-119.)

Hand Signals

Before any lifts are commenced, the crane operator, signalers, riggers, and all other parties involved must re-familiarize themselves with appropriate hand signals, as signals vary from job to job and region to region. Everyone should be familiar with hand signals as outlined in ANSI/ASME B30.5, *Mobile and Locomotive Cranes*. (See Figures 25-1A, 25-1B, and 25-1C.) OSHA 1926.550(a)(4) states: "Hand signals to crane and derrick operators shall be those prescribed by the applicable ANSI standard for the type of crane in use. An illustration of the signals shall be posted at the job site."

Signaling Devices

On lifts where signalers are outside the direct view of the operator due to elevation or in blind areas, either a telephone or radio is a necessity. Agreement should be reached in pre-construction planning as to needed communication systems. The job hazard analysis should also cover this. (See Chapter 11, Vision.)

Lifting Capabilities

During pre-construction planning, lifting capabilities of the cranes to be used should by analyzed by an engineer competent in this field to establish whether the boom length or lifting capacity is sufficient for the intended radius.

Unknown Loads

When the weight of a load to be lifted is unknown, it should be examined and calculated by a registered Professional Engineer who is competent in this regard when the load cannot be weighted. In many instances the actual weight of a load can be determined with a crane equipped with a load moment device which has a readout of actual weight. Caution must be taken in such lifts not to exceed the rated capacity for the crane at the particular lifting radius.

Rigging Practices

Planning should also see that all rigging practices conform to OSHA 1910.184, "Slings" and OSHA 1926.251, "Rigging Equipment for Material Handling."

Figure 25-1A **Hand Signals**

HOIST. With forearm vertical, forefinger pointing up, move hand in small horizontal circle.

LOWER. With arm extended downward, forefinger pointing down, move hand in small horizontal circle.

USE MAIN HOIST. Tap fist on head; then use regular signals.

USE WHIPLINE (Auxiliary Hoist). Tap elbow with one hand; then use regular signals.

RAISE BOOM. Arm extended, fingers closed, thumb pointing upward.

LOWER BOOM. Arm extended, fingers closed, thumb pointing downward.

MOVE SLOWLY. Use one hand to give any motion signal and place other hand motionless in front of hand giving the motion signal. (Hoist slowly shown as example.)

RAISE THE BOOM AND LOWER THE LOAD. With arm extended, thumb pointing up, flex fingers in and out as long as load movement is desired.

LOWER THE BOOM AND RAISE THE LOAD. With arm extended, thumb pointing down, flex fingers in and out as long as load movement is desired.

Figure 25-1B **Hand Signals**

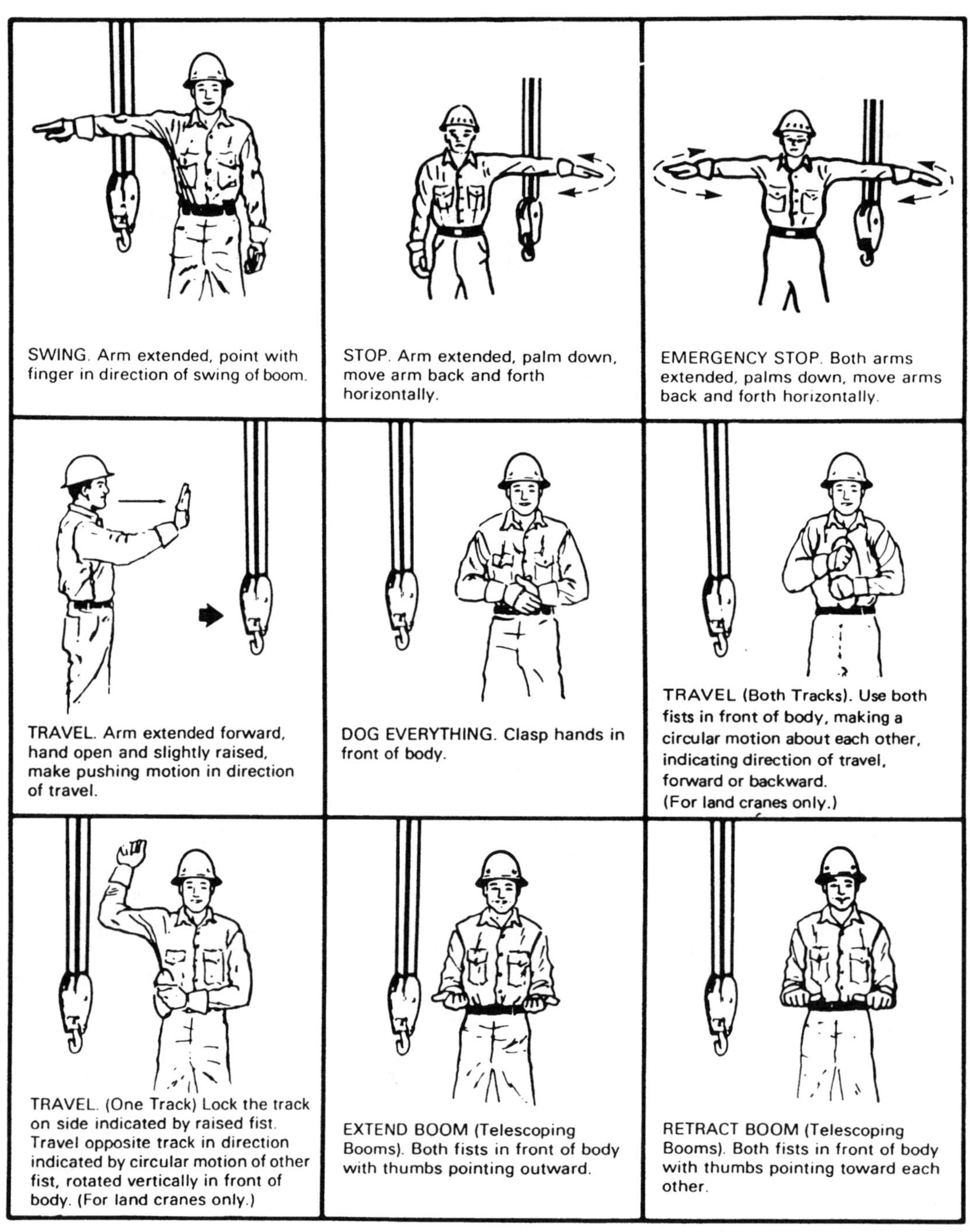

These twenty illustrations are courtesy of ASME, from ANSI B30.15–1973

Controlling the Load

Another important topic for the planning meeting is the use of tag lines to control movement of the load. Normally, when hoisting a load, the lay or twist of wire rope causes rotation when the load becomes suspended. OSHA 1910.180(h)(3)(xvi) states: "… A tag or restraint line shall be used when rotation of the load is hazardous."

Wire Rope Requirements

The planning meeting should also assure that the wire rope used for various hoist lines or pendants meets the crane manufacturer's recommendations.

Annual Inspections

Current crane inspection certificates consistent with OSHA 1910.179(j), 1910.180(d), 1910.181(d), Marine Terminal Standard 1917.45(k), and OSHA 1926.550(a)(6) should be available for all cranes used on the job site. There are a number of firms who perform these inspections.

Preventive Maintenance

All cranes must have continual servicing and preventive maintenance performed. Planning should require that records be checked to determine if current maintenance has been performed and is consistent with the crane manufacturer's recommendations and the crane is fit for use.

Mobile Crane Hazard Analysis Worksheet

1. Review proposed facility design drawings to determine the types of cranes that may be used to accomplish the erection of proposed structure. Assure that cranes will not be used adjacent to powerlines, that safe clearance from roadway traffic exists, that there is sufficient space for outriggers to be extended, that there is adequate shoring to support a crane if work must be done next to an excavation, and other hazards that might arise.

   Findings and comments:

2. Review the contract and specifications for the proposed project to assure that they specify adequate lifting capacity, appropriate safeguards such as load-moment indicators, anti-two-blocking devices, proximity alarms, insulated links, communication systems if blind lifts are anticipated, and other hazard prevention measures. Assure that they also require current certificates of crane inspection, responsible supervision of crane use, and competent and physically qualified crane operators.

   Findings and comments:

3. Review the process for qualifying competent and careful contractors who have a good record of safe crane operation.

   Findings and comments:

4. List all crane hazards and preventive measures discussed at pre-construction safety conferences conducted prior to the issuance of the notice to proceed.

Findings and comments:

5. Review the project safety plan of every contractor to be sure it adequately addresses crane operation and necessary crane hazard prevention measures.

Findings and comments:

6. Examine and document each crane operator and his/her qualifications before giving authorization to operate cranes on the project, to include but not be limited to:
   a. Summary of crane operating experience.
   b. List of any previous crane accidents.
   c. Education.
   d. Current doctor's physical examination for:
      1) Vision with a minimum of 20/30 Snellen in one eye and 20/50 in the other, with or without corrective lenses; and the ability to distinguish colors.
      2) Adequate hearing, with or without a hearing aid.
      3) Sufficient strength, endurance, agility, coordination, and speed of reaction to meet equipment needs.
      4) No evidence of physical defects or emotional instability that could interfere with performance.
      5) No proneness to seizures or loss of physical control.
      6) No evidence of drug or alcohol use.
   e. Examination of operator's demonstrated ability:
      1) Demonstrating the setting up of the crane, to include the use of outriggers.
      2) Demonstrating the ability to attach or detach counterweights if the crane has removable or adjustable counterweights.
      3) Demonstrating a knowledge of the function of each control and accessory.
      4) Demonstrating the ability to handle a load, including lifting with the hoist line, raising and lowering of the boom, extending and retracting boom if hydraulic, and sluing.
      5) Demonstrating the ability to map the danger zone as outlined in Chapter 4, Powerline Contact, and to be able to position the crane where it will not intrude into the danger zone at any boom radius.
      6) Answering questions to determine operator's understanding of the methods of assembling or disassembling the jib or boom.

    7) Answering questions to determine if operator knows to whom to report when the crane cannot be operated safely because the load exceeds the rated capacity, when the powerline danger zone will be entered if the lift is made, when adequate communication for blind lifts is needed, etc.

f. The passing of a written test to determine the ability to read and interpret load charts on all cranes to which assigned, including calculations for a maximum load for several given radii and the use of the jib. See *Mobile Crane Manual* by D. E. Dickie (Construction Safety Association of Ontario, 1982). In the Appendix of this book there are representative exercises for the various cranes manufactured in the United States, with questions and answers about proper calculations.

    Findings and comments:

7. Before authorizing its use, examine the crane itself:
   a. Does it have a current certification?
   b. Conduct a load test by maneuvering a specified test load through maximum lift height, lift radius, and boom quadrant. Such tests shall not exceed the manufacturer's rated capacity for boom angle radius. A log of this quality test will be retained.
   c. Is a load-moment limiter (LML) or load-moment indicator (LMI) installed and functional?
   d. Are outriggers functional with adequate pads?
   e. Is an anti-two-blocking device installed and functional?
   f. Is the boom-hoist kickout functional if the crane has a latticework boom?
   g. Does the latticework boom have functional boom stops?
   h. Is there adequate clearance between cab/carrier and cab/fixed object to avoid this critical pinch point? Are there conspicuous barriers and warning to exclude people from the danger zone?
   i. Is there adequate guarding of moving parts inside of crane cab?
   j. Are there safety latches on hoisting hooks?
   k. Is there a functioning communication system (telephone or radio) for blind lifts?
   l. Is there a functioning backup and/or travel alarm?
   m. Inspect cable of hoist line, fly line, boom hoist cable, and condition of all sheaves.
   n. Does the crane have safe access?
   o. Does the crane have adequate ventilation and heating in the operator's compartment?
   p. Is the crane equipped with a seat belt and crush-resistant cab to protect the operator in the event of upset?
   q. Thoroughly examine crane for other hazards and necessary safety measures.

    Findings and comments:

8. Inspect the work site for:
    a. Powerlines.
    b. Adequate footings, not endangered by excavations or unstable soil conditions.
    c. Sufficient space for outriggers so they will not intrude into roadways and other access.
    d. Barricading of the danger zone created by the rotating cab.

    Findings and Comments:

9. Determine if crane operators, riggers, supervisors, and others who are required to work with the crane have a mutual understanding and knowledge of the use of appropriate hand signals. Determine if the crew knows who is assigned to provide signals for the crane operator. There should be only one person responsible for this. The signals and the communication system should be agreed upon by all persons involved. Determine whether the radio, telephone, or other means of communication for blind lifts is functioning.

    Findings and Comments:

10. Determine if the crane can be locked and whether its use is controlled by permit to avoid unauthorized use.

11. In 1995, a crane operator's certification program was developed by the National Commission for the Certification of Crane Operators, a not-for-profit organization funded by industry. For additional information, write or call:

     National Commission for the Certification of Crane Operators
     2750 Prosperity Avenue, Suite 505
     Fairfax, Virginia 22031-4312
     Telephone: (703) 560-2391
     Fax: (703) 560-2392

Bibliography

OSHA Requirements

1. *Occupational Safety and Health Act of 1970*, Public Law 91-596.
2. *Code of Federal Regulations*, General Industry Standards, Title 29, Part 1910.
3. *Code of Federal Regulations*, Construction Standards, Title 29, Part 1926.
4. *Safety and Health Regulations for Marine Terminals*, Title 29, Part 1917.
5. *Safety and Health Regulations for Longshoring*, Title 29, Part 1918.
6. *Concepts and Techniques of Machine Guarding*, OSHA No. 3067, 1980.

ANSI Standards

Safety Standards for Cableways, Cranes, Derricks, Hoists, Hooks, Jacks, and Slings

7. ANSI B30.1 — Jacks
8. ANSI B30.2 — Overhead and Gantry Cranes
9. ANSI B30.3 — Hammerhead Tower Cranes
10. ANSI B30.4 — Portal, Tower, and Pillar Cranes
11. ANSI B30.5 — Mobile and Locomotive Cranes
12. ANSI B30.6 — Derricks
13. ANSI B30.7 — Base Mounted Drum Hoists
14. ANSI B30.8 — Floating Cranes and Floating Derricks
15. ANSI B30.9 — Slings
16. ANSI B30.10 — Hooks
17. ANSI B30.11 — Monorails and Underhung Cranes
18. ANSI B30.12 — Handling Loads Suspended from Rotorcraft
19. ANSI B30.13 — Storage/Retrieval (S/R) Machines and Associated Equipment
20. ANSI B30.14 — Side Boom Tractors
21. ANSI B30.15 — Mobile Hydraulic Cranes (Note: B30.15-1973 has been withdrawn. The revision of B30.15 is included in the latest edition of B30.5.)
22. ANSI B30.16 — Overhead Hoists
23. ANSI B30.17 — Overhead and Gantry Cranes
24. ANSI B30.18 — Stacker Cranes
25. ANSI B30.19 — Cableways
26. ANSI B30.20 — Below-the-Hook Lifting Devices
27. ANSI B30.21 — Manually Lever Operated Hoists (Note: This standard is in the developmental stage.)
28. ANSI B30.22 — Articulating Boom Cranes

Other ANSI Standards

29. ANSI A12.1 — Safety Requirements for Floor and Wall Openings, Railings, and Toeboards
30. ANSI A14.3 — Safety Requirements for Fixed Ladders

31. ANSI B15.1   Safety Standard for Mechanical Power Transmission Apparatus
32. ANSI C2   National Electrical Safety Code
33. ANSI/IEEE C2   National Electrical Safety Code Interpretations, 1961-1977 Inclusive
34. ANSI/IEEE C2   National Electrical Safety Code Interpretations, 1978-1980 Inclusive, and Interpretations Prior to the 6th Edition, 1961
35. ANSI Z35.1   Specifications for Accident Prevention Signs

SAE

SAE Handbook, SAE Recommended Practices, Volume 4, Society of Automotive Engineers

36. SAE J115   Safety Signs
37. SAE J159   Crane Load Moment System
38. SAE J185   Access Systems for Off-Road Machines
39. SAE J220   Crane Boomstop
40. SAE J375   Radius-of-Load and Boom Angle Measuring System
41. SAE J376   Load Indicating Devices in Lifting Crane Service
42. SAE J765   Crane Load Stability Test Code
43. SAE J820   Crane Hoist Line Speed and Power Test Code
44. SAE J881   Lifting Crane Sheave and Drum Sizes
45. SAE J959   Lifting Crane, Wire-Rope Strength Factors
46. SAE J983   Crane and Cable Excavator Basic Operating Control Arrangements
47. SAE J987   Crane Structures — Method of Test
48. SAE J999   Crane Boom Hoist Disengaging Device
49. SAE J1028   Mobile Crane Working Area Definitions
50. SAE J1040c   Performance Criteria for Rollover Protective Structures (ROPS) for Construction, Earthmoving, Forestry, and Mining Machines
51. SAE J1063   Cantilevered Boom Crane Structures — Method of Test
52. SAE J1180   Telescoping Boom Length Indicating System
53. SAE J1238   Rating Lift Cranes on Fixed Platforms Operating
54. SAE J1257   Rating Chart for Cantilevered Boom Cranes
55. SAE J1289   Mobile Crane Stability Ratings
56. SAE J1332   Rope Drum Rotation Indicating Device
57. *SAE Standard J2270 Surface Vehicle Standard*, "Ships Systems & Equipment, Threaded Fasteners—Inspection, Test, and Installation Requirements," January 1996.
58. *SAE Standard J429*, "Mechanical and Material Requirements for Externally Threaded Fasteners," January 1980.
59. *SAE Standard J1199*, "Mechanical and Material Requirements for Metric Externally Threaded Steel Fasteners," February 1978.
60. *SAE Standard J1216*, "Test Methods for Metric Threaded Fasteners," March 1978.
61. *SAE Standard J82*, "Mechanical and Quality Requirements for Machine Screws," June 1979.
62. *SAE Standard J995*, "Mechanical and Material Requirements for Steel Nuts," June 1979.
63. *SAE Standard J1102*, "Mechanical and Material Requirements for Wheel Bolts," October 1974.
64. *SAE Recommended Practice J430*, "Mechanical and Chemical Requirements for Non-threaded Fasteners (Rivets)," October 1968.
65. *SAE Recommended Practice J933*, "Mechanical and Quality Requirements for Tapping Screws" June 1979.
66. *SAE Standard J78*, "Steel Self-Drilling Tapping Screws," June 1979.
67. *SAE Standard J81*, "Thread Rolling Screws," June 1979.
68. *SAE Standard J1237*, "Metric Thread Rolling Screws," May 1979.

69. *SAE Recommended Practice J1061a*, "Surface Discontinuities on General Application Bolts, Screws, and Studs," July 1975.
70. *SAE Recommended Practice J123c*, "Surface Discontinuities on Bolts, Screws, and Studs," June 1976.
71. *SAE Recommended Practice J122a*, "Surface Discontinuities on Nuts," July 1975.
72. *SAE Recommended Practice J121a*, "Decarburization in Hardened and Tempered Threaded Fasteners," July 1976.

Other Requirements

73. *National Electrical Code*, National Fire Protection Association (NFPA), Chapter 6, Article 610, Cranes and Hoists.
74. *Safety and Health Requirements Manual*, EM 385-1-1, U.S. Army Corps of Engineers, Department of the Army.
75. *Mobile Power Crane and Excavator Standards*, PCSA Standard No. 1, Power Crane and Shovel Association (PCSA), a Bureau of the Construction Manufacturer's Association.
76. *Mobile Hydraulic Crane Standards*, PCSA Standard No. 2, PCSA.
77. *Hydraulic Excavators and Telescoping Boom Cranes*, PCSA.
78. *Cable Controlled Power Cranes, Draglines, Hoes, Shovels, and Clamshells*, PCSA.

Other References

79. Dickie, D. E. 1975. *Crane Handbook*. Construction Safety Association of Ontario, 74 Victoria Street, Toronto, Ontario, Canada M5C2A5, (416) 336-1501.
80. Dickie, D. E. 1982. *Mobile Crane Manual*. Construction Safety Association of Ontario.
81. Dickie, D. E. 1975. *Rigging Manual*. Construction Safety Association of Ontario,
82. MacCollum, D. V. 1984. Crane Design Hazard Analysis, Chapter 8. *Automotive Engineering and Litigation*, Volume 1. Peters, G. A., and Peters, B. J., eds.
83. Engineering and Technology *Accident Prevention Manual for Industrial Operations*, Ninth Edition 1988. Chapters 4 and 5. National Safety Council.
84. Administration and Programs, *Accident Prevention Manual for Industrial Operations*, Ninth Edition 1988. National Safety Council. pp. 189-194.
85. U.S. Army Corps of Engineers, *Safety and Health Requirements Manual*, EM 385-1-1 (revised October 1992), Section 16, Machinery and Mechanized Equipment.
86. Shapiro, H. I. 1980. *Cranes and Derricks*. New York: McGraw-Hill, Inc.
87. Shapiro, H. I., Shapiro, J. P., and Shapiro, L. K. 1991. *Cranes and Derricks*, 2nd Edition. New York: McGraw-Hill, Inc.
88. Kogan, J. 1976. *Crane Design, Theory and Calculations of Reliability*. New York: John Wiley & Sons.
89. Campbell, D. H. 1974. *Craning Principles*. Construction Management Group, Department of Civil Engineering, University of Waterloo, Waterloo, Ontario, Canada.
90. The Royal Society for the Prevention of Accidents. 1967. *A Safety Handbook for Mobile Cranes*. Terminal House, 52 Grosvenor Gardens, London, S.W.1.
91. The Royal Society for the Prevention of Accidents. 1974. *A Safety Handbook for Mobile Cranes*, Cannon House, The Priory, Queensway, Birmingham B46BS.
92. Campbell, D. H. 1988. *Mobile Craning Today*. International Union of Operating Engineers, Local 793, Ontario, Canada.
93. Kruger & Co., K.G. 1977. *Safe Working with Truck-Mounted Cranes*. Hamburg, Germany: Berufsgenossenschaft fur Fahrzeughaltungen.
94. Rothbart, Harold (ed.) *Mechanical Design Handbook*. 1995. New York: McGraw-Hill, Inc.
95. Beyer, Howard. *Selection of Materials for Component Design*. American Society for Metals.
96. Avallone, Eugene A. and Baumeister III, Theodore (eds.). 1996. *Marks' Standard Handbook foir Mechanical Engineers,10th Ed.* New York: McGraw Hill, Inc.
97. Parmley, Robert O. (ed.) 1997. *Standard Handbook of Fastening and Joining, 3rd Ed.* New York: McGraw Hill, Inc.

98. Bickford, John H. 1990. *An Introduction to the Design and Behavior of Bolted Joints, 3rd Ed.* New York: Marcel Dekker, Inc.

99. *Machine Design Magazine,* "Fastening and Joining Reference Issue," Penton, IPC Cleveland, OH. November 17, 1983.

100. Oberg, Erik and McCauley, Christopher J. (eds.). 1996. "Bolts, Nuts, Screws, Washers, Nails & Spikes," *Machinery's Handbook,* 25th Revised Edition, New York: Industrial Press, Inc.

Human Factors

101. Woodson, W. E., and Conover, D. W. 1964. *Human Engineering Guide for Equipment Designers.* University of California Press.

102. Meister, David and Gerald F. Rabideau. 1965. *Human Factors Evaluation in System Development.* John Wiley & Sons, Inc.,

103. Chapanis, A. 1965. *Man-Machine Engineering.* Brooks/Cole Publishing Company,.

104. Fitts, P. M., and Posner, M. I. 1967. *Human Performance.* Brooks/Cole Publishing Company.

105. Surry, J. 1971. *Industrial Accident Research, A Human Engineering Appraisal.* University of Toronto, Labour Safety Council.

106. Van Cottand, H. P., and Kinkade, R.G. *Human Engineering Guide to Equipment Design.* Army-Navy-Air Force Steering Committee.

107. Huchingson, R. D. 1981. *New Horizons for Human Factors in Design.* McGraw-Hill Book Company.

Pre-Job Planning

108. Ely, M. D. 1960. "Coordinating the Safety Program with the Independent Contractors." Paper presented in 1960 at the National Safety Council's National Safety Congress.

109. Ely, M. D. 1960. "It's Part of the Contract." *National Safety News.*

110. *Accident Prevention Manual for Industrial Operations,* 6th Edition. 1969. Chicago: National Safety Council.

111. Holmes, H. J. 1969. "What is CPM?" *National Safety Congress Transactions,* Volume 8. Chicago: National Safety Council, p.5.

112. Jenkins, R. L. 1971. *Contractor's Management Handbook.* Chapter 11.

113. Klashak, A. R. "Contract Specifications: Another Dimension to Safety." *Safety Newsletter,* August 1976.

114. Levitt, R. E., and Parker, H. W. September, 1976. "Reducing Construction Accidents—Top Management's Role." *Journal of the Construction Division,* Chicago: National Safety Council.

115. "Construction Control." 1978. *The BOCA Basic Building Code.*

116. Levitt, R. E., Parker, H. W., and Samelson, N. M. August, 1981. *Improving Construction Safety Performance: The User's Role.* Technical Report No. 260. Department of Civil Engineering, Stanford University.

117. MacCollum, David V., P.E., CSP. 1995. *Construction Safety Planning.* New York: John Wiley & Sons, Inc.

118. Samelson, N. M., Levitt, R. E., and Parker, H. W. 1982. *The Role of Owners in Reducing Construction Accident Costs.* Department of Civil Engineering, Stanford University.

119. Samelson, N. M., and Levitt, R. E. April 26-30, 1982. "Owner's Guidelines for Selecting Safe Contractors." Preprint 82-040. Proceedings of American Society of Civil Engineers.

Foreign Regulations

120. "Avoidance of Danger from Overhead Electric Lines." *Guidance Note General Series 6.* April, 1977. HM Factory Inspectorate of the Health and Safety Executive and the Electricity Boards of England, Wales and Scotland.

121. United Kingdom Regulation 44 of the Construction (General Provisions) Regulations of the Factories Act of 1961.

122. Indications of Safe Working Load of Jib Cranes. *Factories*. United Kingdom Regulation 1581 of the Construction (Lifting Operations) Regulations, 1961, Part II, 30.

123. Netherlands Normalization Institute (NNI), *NEN Norm Standards 2017-2028*.

Magazines

124. *Cranes Today* and *Lift Equipment* are excellent reference magazines. Subscriptions to these magazines are usually free. To obtain a subscription, contact: *Cranes Today*, MBC, Audit House, Field End Road, Eastcote, Ruislip, Middlesex, HA4 9BR, UK; and *Lift Equipment*, 10229 E. Independence Avenue, Independence, Missouri 64053.

System Safety

125. U.S. Air Force Systems Command/National Aeronautics & Space Administration. 1969 to Present. *System Safety, Design Handbooks Series*, 2nd Edition.

126. MacCollum, D. V. Reliability as a Quantitative Safety Factor. *Journal of the American Society of Safety Engineers*, May, 1969.

127. Hammer, W. 1972. *Handbook of System & Product Safety*. Prentice-Hall.

128. Johnson, William G., and Grandjean, E. 1973. *The Management Oversight & Risk Tree — MORT*. Superintendent of Documents.

129. Rogers, W. P. 1980. *Introduction to System Safety Engineering*. Robert E. Krieger Publishing Co.

130. Hammer, W. 1993. *Product Safety Management and Engineering*. Des Plaines, Illinois: American Society of Safety Engineers.

131. Hammer, W. 1981. *Occupational Safety Management and Engineering*. Prentice-Hall.

132. *Military Standard, System Safety Program Requirements*, MIL-STD-882B, March 30, 1984, Superseding MIL-STD-882A, June 28, 1977.

133. *System Safety Engineering and Management*, Army Regulation 385-16, September 3, 1985.

Glossary

Accessory
A secondary part or assembly of parts which contributes to the overall function and usefulness of a machine.

Angle Indicator
Accessory which measures the angle of the boom to the horizontal.

ANSI
American National Standards Institute.

Anti-Two-Blocking Device
A sensing system to prevent two-blocking of the hoist hook assembly with the boom tip.

Articulated Boom
Boom that is either able to extend with a scissors movement or bends inwardly.

Automatic Crane
Crane which when activated operates through a preset cycle or cycles.

Auxiliary Hoist
Supplemental hoisting unit of lighter capacity and usually higher speed than provided for the main hoist.

Axis of Rotation
Vertical axis around which the crane superstructure rotates.

Axle
The shaft or spindle with which or about which a wheel rotates. On truck-and wheel-mounted cranes it refers to an automotive type of axle assembly including housings, gearing, differential, bearings, and mounting appurtenances.

Axle (bogie)
Two or more automotive-type axles mounted in tandem in a frame so as to divide the load between the axles and permit vertical oscillation of the wheels.

Base (mounting)
Traveling base or carrier on which the rotating superstructure is mounted such as a car, truck, crawler, or wheel platform.

Boom
Member hinged to the front of the rotating superstructure with the outer end supported by ropes leading to a gantry or A-frame and used for supporting the hoisting tackle.

Boom Angle
Angle between the longitudinal centerline of the boom and the horizontal. The boom longitudinal centerline is a straight line between the boom foot pin (heel pin) centerline and boom point sheave pin centerline.

Boom Hoist
Hoist drum and rope reeving system used to raise and lower the boom. The rope system may be all live reeving or a combination of live reeving and pendants.

Boom Stop
Device used to limit the angle of the boom at the highest position.

Brake
Device used for retarding or stopping motion by friction or power means.

Bridge

That part of a crane consisting of girders, trucks, end ties, footwalks, and the drive mechanism which carries the trolley or trolleys.

Bridge Travel

Crane movement in a direction parallel to the crane runway.

Bumper (buffer)

Energy absorbing device for reducing impact when a moving crane or trolley reaches the end of its permitted travel, or when two moving cranes or trolleys come in contact.

Cab

Housing which covers the rotating superstructure machinery and/or operator's station. On truck-crane trucks, a separate cab covers the driver's station.

Cab-Operated Crane

Crane controlled by an operator in a cab located on the bridge or trolley.

Cantilever Boom

Boom that is extendible, usually hydraulically, and is self-supporting.

Cantilever Gantry Crane

Gantry or semi-gantry crane in which the bridge girders or trusses extend transversely beyond the crane runway on one or both sides.

Clearance

Distance from any part of the crane to a point of the nearest obstruction.

Clutch

A friction, electromagnetic, hydraulic, pneumatic, or positive mechanical device for engagement or disengagement of power.

Collectors, Current

Contacting devices for collecting current from runway or bridge conductors.

Conductors, Bridge

Electrical conductors located along the bridge structure of a crane to provide power to the trolley.

Conductors, Runway (main)

Electrical conductors located along a crane runway to provide power to the crane.

Control Braking

Method of controlling crane motor speed when in an overhauling condition.

Countertorque

Method of control by which the power to the motor is reversed to develop torque in the opposite direction.

Counterweight

Device used to supplement the weight of the machine in providing stability for lifting working loads.

Crane

Machine used for lifting and lowering a load and moving it horizontally, with the hoisting mechanism an integral part of the machine. Cranes, whether fixed or mobile, are driven manually or by power.

Crawler Crane

Consists of a rotating superstructure with power plant, operating machinery, and a boom, mounted on a base, equipped with crawler treads for travel. Its function is to hoist and swing loads at various radii.

Designated

Selected or assigned by the employer or the employer's representative as being qualified to perform specific duties.

Drag Brake

Brake which provides retarding force without external control.

Drift Point

Point on a travel motion controller which releases the brake while the motor is not energized. This allows for coasting before the brake is set.

Drum

Cylindrical members around which ropes are wound for raising and lowering the load or boom.

Dynamic

Method of controlling crane motor speeds when in the overhauling condition to provide a retarding force.

Dynamic (loading)

Loads introduced into the machine or its components by forces in motion.

Emergency Stop Switch

A manually or automatically operated electric switch to cut off electric power independently of the regular operating controls.

Exposed

Capable of being contacted inadvertently. Applied to hazardous objects not adequately guarded or isolated.

Fail-Safe

A provision designed to automatically stop or safely control any motion in which a malfunction occurs.

Floor-Operated Crane

Crane which is pendant or non-conductive rope controlled by an operator on the floor or an independent platform.

Footwalk

Walkway with handrail, attached to the bridge or trolley, for access purposes.

Free Fall

The load hook can be reeled out without any motor speed.

Gantry (A-frame)

Structural frame, extending above the superstructure, to which the boom support ropes are reeved.

Gantry Crane

Crane similar to an overhead crane except that the bridge for carrying the trolley or trolleys is rigidly supported on two or more legs running on fixed rails or other runway.

Holding Brake

Brake that automatically prevents motion when power is off.

Hoist

Apparatus which may be a part of a crane, exerting a force for lifting or lowering.

Hoist Chain

Load bearing chain in a hoist.

Hoist Motion

That motion of a crane which raises and lowers a load.

Hot Metal Handling Crane

Overhead crane used for transporting or pouring molten material.

Hydraulic Boom

Boom that is extended with hydraulic rams.

Jib

Extension attached to the boom point to provide added boom length for lifting specified loads. The jib may be in line with the boom or offset to various angles.

Latticework Boom

Boom that is a fabricated truss.

Limit Switch

A switch which is operated by some part or motion of a power-driven machine or equipment to alter the electric circuit associated with the machine or equipment.

Load

The total superimposed weight on the load block or hook.

Load (working)

The external load, in pounds, applied to the crane, including the weight of load-attaching equipment such as load blocks, shackles, and slings.

Load Block (lower)

The assembly of hook or shackle, swivel, sheaves, pins, and frame suspended by the hoisting ropes.

Load Block (upper)

The assembly of hook or shackle, swivel, sheaves, pins, and frame suspended from the boom point.

Load Hoist

Hoist drum and rope reeving system used for hoisting and lowering loads.

Load Ratings

Crane ratings in pounds established by the manufacturer.

Locomotive Crane

Consists of a rotating superstructure with power plant, operating machinery, and a boom, mounted on a base or car equipped for travel on railroad track. It may be self-propelled by an outside source. Its function is to hoist and swing loads at various radii.

Magnet

Electromagnetic device carried on a crane hook to pick up loads magnetically.

Main Hoist

Hoist mechanism provided for lifting the maximum rated load.

Man Trolley

Trolley having an operator's cab attached thereto.

Main Switch

A switch controlling the entire power supply to the crane.

Master Switch

A switch which dominates the operation of contractors, relays, or other remotely-operated devices.

Outriggers

Extendible or fixed metal arms, attached to the mounting base, which rest on supports at the outer ends.

Overhead Crane

Crane with a movable bridge carrying a movable or fixed hoisting mechanism and traveling on an overhead fixed runway structure.

Pin-Up Guys/Pendant Guys

Cables used to support a latticework boom.

Power Operated Crane

Crane whose mechanism is driven by electric, air, hydraulic, or internal combustion means.

Pulpit-Operated Crane

Crane operated from a fixed operator station not attached to the crane.

Rail Clamp

A tong-like metal device, mounted on a locomotive crane car, which can be connected to the track.

Rated Load

Maximum load for which a crane or individual hoist is designed and built by the manufacturer and shown on the equipment nameplate(s).

Reeving

A rope system unless otherwise specified.

Regenerative

Form of dynamic braking in which the electrical energy generated is fed back into the power system.

Remote-Operated Crane

Crane controlled by an operator not in a pulpit or in the cab attached to the crane, by any method other than pendant or rope control.

Rope

Refers to a wire rope unless otherwise specified.

Rough-Terrain Crane

A four-wheel drive crane, steer operated from the crane cab. Its function is to hoist and swing loads at various radii.

Running Sheave

A sheave which rotates as the load block is raised or lowered.

Runway

An assembly of rails, beams, girders, brackets, and framework on which the crane or trolley travels.

Semi-gantry Crane

Gantry crane with one end of the bridge rigidly supported on one or more legs that run on a fixed rail or runway, the other end of the bridge being supported by a truck running on an elevated rail or runway.

Sheave or Pulley
Allows the wire rope to move over an angle.

Side Loading
A load applied at an angle to the vertical plane of the boom.

Side Pull
That portion of the hoist pull acting horizontally when the hoist lines are not operated vertically.

Sluing
Rotation of the crane cab to allow the boom to revolve 360 degrees around the center pin.

Span
The horizontal distance center to center of runway rails on bridge cranes.

Standby Crane
Crane that is not in regular service but which is used occasionally or intermittently as required.

Standing (guy) Rope
Supporting rope which maintains a constant distance between the points of attachment to the two components connected by the rope.

Storage Bridge Crane
Gantry type crane of long span, usually used for bulk storage of material. The bridge girders or trusses are rigidly or non-rigidly supported on one or more legs. It may have one or more fixed or hinged cantilever ends.

Stop
Device to limit travel of a trolley or crane bridge. This device normally is attached to a fixed structure and normally does not have energy absorbing ability.

Structural Competence
Ability of the machine and its components to withstand the stresses imposed by applied loads.

Superstructure
Rotating upper frame structure of the machine and the operating machinery mounted thereon.

Swing
Rotation of the superstructure for movement of loads in a horizontal direction about the axis of rotation.

Swing Mechanism
Machinery involved in providing rotation of the superstructure.

Switch
Device for making, breaking, or for changing the connections in an electric circuit.

Tackle
Assembly of ropes and sheaves arranged for hoisting and pulling.

Telescoping Boom
Boom which is in several sections that is hydraulically extended or retracted.

Travel
Function of the machine moving from one location to another, on a job-site.

Travel Mechanism
Machinery involved in providing travel.

Trolley
Unit which travels on the bridge rails and carries the hoisting mechanism.

Trolley Boom
A fixed-length boom where the hook is cable-moved back and forth along the boom.

Trolley Travel
Trolley movement at right angles to the crane runway.

Truck
The unit consisting of a frame, wheels, bearings, and axles which supports the bridge girders and trolleys

Truck Crane
Consists of a rotating superstructure with power plant, operating machinery, and a boom, mounted on an automotive truck equipped with a power plant for travel. Its function is to hoist and swing loads at various radii.

Two-Blocking
The hoist line hook assembly of either the headache ball or sheaves touch the boom tip, often resulting in a parting of the hoist cable or damage to the boom tip and/or sheave assembly.

Wall Crane
Crane having a jib with or without trolley and supported from a side wall or line of columns of a building. It is a traveling type and operates on a runway attached to the side wall or columns.

Wheelbase
Distance between centers of front and rear axles. For a multiple axle assembly the axle center for wheel base measurement is taken as the midpoint of the assembly.

Wheel-Mounted Crane
Consists of a rotating superstructure with power plant, operating machinery and boom, mounted on a base or platform equipped with axles and rubber-tired wheels for travel. The base is usually propelled by the engine in the superstructure, but it may be equipped with a separate engine controlled from the superstructure. Its function is to hoist and swing loads at various radii.

Whipline (auxiliary hoist)
A separate hoist rope system of lighter load capacity and higher speed than provided by the main hoist.

Winch Head
Power driven spool for handling of loads by means of friction between fiber or wire rope and spool.

Index

A

A-frames. *See:* Latticework booms

Acceptable risk, 8

Access
- pinch points and, 63
- unsafe, 59-62

Aerial lift(s)
- and aerial lift platforms, 20
- boom conductivity of, 32
- retrofit kits for, 33
- unintended movement of, 76
- upset of self-propelled, without counterweights, 116

Alarms. *See:* Warning systems

American Trial Lawyers Association (ATLA), 32

Anchor systems, 131-136
- capacity requirements for each zone in, 135
- catenary method for pre-tensioning guywires in, 133, 136
- design of, 133
- four basic types of manufactured anchors for, 132
- guywire use in, 132
- mast-type devices for, 131
- recommended anchor locations for, 134-135

Annual inspections, 151

ANSI Standard
- A12.1, Safety Requirements for Floor and Wall Openings, Railings, and Toeboards, 61
- A14.3, Safety Requirements for Fixed Ladders, 60-61
- A92.2, Vehicle-Mounted Elevating and Rotating Aerial Devices, 33, 117
- A92.2-2001, Vehicle-Mounted Elevating and Rotating Aerial Devices, 44
- A92.2, Vehicle-Mounted Elevating and Rotating Work Platforms, 74
- A92.5, Boom-Supported Elevating Work Platforms, 76
- A92.5, Section 5.2.4, Outriggers, Stabilizers, and Extendible Axles, 44
- A92.5, Section 6, Boom-Supported Elevating Work Platforms, 117
- A92.5, Boom-Supported Elevating Work Platforms, 13.2.2 (1), 33
- A92.6, Self-Propelled Elevating Work Platforms, 14.2.2, 33
- A92.6-1979, 45
- A92.6-1979, Section 6.2.5, Self-Propelled Elevating Work Platforms, 76
- B15.1, Safety Standard for Mechanical Power Transmission Apparatus, 70
- B15.1-1992 revised, Safety Standard for Mechanical Power Transmission Apparatus, 70
- B30.11, Monorail Systems and Underhung Cranes, Chapter 11-1, Construction and Installation, 129
- B30.15, Mobile Hydraulic Cranes, 41, 92
- B30.15-1975, Mobile Hydraulic Cranes, 41, 48
- B30.15, Section 15-3.4.2, 24
- B30.2.0, Overhead and Gantry Cranes, 61
- B30.2, Overhead and Gantry Cranes, 80
- B30.2.0, Overhead and Gantry Cranes, Section 2-1.10.6, Runway Conductors, 125
- B30.2.0, Overhead and Gantry Cranes, Section 2-2.3.2, 126
- B30.2.0-1976, Warning Devices, 82

B30.22, Articulating Boom Cranes, Section 22-1.6.1(c), 74
B30.22, Section 22-3.3, 24
B30.3-1975, Hammerhead Tower Cranes, 80
B30.3, Hammerhead Tower Cranes, Section 3-1.1.1.c, 123
B30.3, Hammerhead Tower Cranes, Section 3-1.2.16, 137
B30.4-1973, Portal, Tower and Pillar Cranes, 80
B30.5, Mobile and Locomotive Cranes, 41, 64
B30.5, Mobile and Locomotive Cranes, 80
B30.5-1989, 48
B30.5 (Addenda to) in Section 5-1.7.4(b) and (c), 89
B30.5 (Addenda to) Section 5-1.9.10(c), Load Indicators, 37
B30.5-1982, Mobile and Locomotive Cranes, Sections 5-1.9.1(a) and (d), 104
B30.5, Section 5-3.4.5, 24
B30.6-1969, Derricks, 133
C2, Table 232, 24
Z26.1-1977, 58
Z26.1a-1980, 58
Z535.1-1991, Safety Color Code, 9
Z535.2-1991, Environmental and Facility Safety Signs, 9
Z535.3-1991, Criteria for Safety Symbols, 9
Z535.4-1991, Product Safety Signs and Labels, 9
Z535.5-1991, Accident Prevention Tags for Temporary Hazards, 9
Anti-two-blocking devices, 47-52. *See also:* Two-blocking
as standard equipment on cranes, 48
foreign vs. U.S. requirements for, 50
patents for, 50-51
retrofit programs for, 52
state requirements for, 50
Articulated boom cranes, 24
requirements to have an emergency stop on, 74
ASA standards
B30.2-1943, 72
B30.2-1943, reaffirmed 1952, Section 1143(a), 82
B30.2-1943, Safety Code for Cranes, Derricks, and Hoists, 97

B

Back-up alarms. *See:* Warning systems
Back-up safety devices
insulated links as, 26
proximity alarms as, 26
Backing, of a crane, 80-81
Baker, O'Neil, 1
Blind lifts, 79-80
considering high lifts as, 79
definition of, 79
use of a radio or telephone for, 79-80
Blind spots. *See:* Vision
Blocking
of latticework boom during disassembly, 106
under outriggers, 41
Boom buckling. *See:* Boom failure
Boom disassembly, 106-110. *See also:* Boom failure *and* Transportation of cranes
on latticework boom cranes, 106-110
statistical data for injuries during, 107
training employees to use blocking during, 106
use of pins during, 106-107
Boom failure, 101-110, 119-120. *See also:* Boom disassembly *and* Boom stops
causes of latticework toppling, 103
repair of damaged booms after, 102
stops to prevent, 103
telescoping booms, 36
use of tag lines to prevent load swinging causing, 101
use of tower crane to avoid, 101
when subjected to side pull, 101
Boom hinge pin assembly failure, 119-120
Boom hoist cable, breaking. *See:* Cables
Boom snub, 103
Boom stops, 103-104. *See also:* Boom failure
Booms, jib. *See:* Jib booms
Bubble indicator, to assist in leveling crane, 105

C

Cab(s)
crush resistant, 8, 53-56
dangers of using, for storage space, 69
operator's vision from, during backing, 80. *See also:* Vision
protecting, from flying or swinging objects, 58

unsafe access to, 59-62
Cable(s)
    binding of, in boom hoist, 88-89
    breaking of boom hoist, 89
    daily inspection of, 86-87
    damage to, caused by sheaves, 85-90
    flattening of (peening), 89
    guarding of sheaves for, 88
    kinking of, 87-88
    latticework booms and, 89
    manufacturer's optimum requirements
      for, 87-88
    manufacturers' service tests on, 88
    number of lifts and lowerings for, 86
    preventing slack in, 87
    proper drum size and, 86
    proper sheave size and, 85-86
    types of failure in, 86
Catenary method. *See:* Anchor systems
Capacity, rated. *See:* Rated capacity
Certification, of crane operators. *See:*
   Operator competency
Chapanis, Alphonse, 2
Communications. *See also:* Blind lifts *and*
   Signals
    absence of, between carrier cab and
      crane cab, 112-13
    closed circuit television use in, 81
    during raising or lowering of tower
      cranes, 121
    two-way radio use in, 79-80, 112
Control boxes. *See:* Powerline contact
Control(s), 47, 71-78
    accessible to activation by outside
      objects, 76-78
    design for dropping the load by free-
      wheeling, 98
    for two-wheel/four-wheel steering, 55
    human performance and, 47, 74, 98
      See also: Human performance
    inadvertent movement of, 55, 71-74
    non-uniform placement of, 71
    on conductive tether, 74
    on track-mounted cranes, 81-82
    reachable from ground, 74
    remote, 24, 74, 99
    two-handed motion for, 72, 76-78
Counterweights, 115-16. *See also:* Pinch
   points
Covers. *See:* Guarding
Crane(s), 13-20
    articulating boom, 15

built-in hazards of, 1. *See also:* Design
   and Hazard(s)
design of, 1-2. *See also:* Design
fixed, 18, 19
general OSHA requirements for, 5. *See*
    also: OSHA Standards
hazards associated with most, 4. *See*
    also: Hazard(s)
hazards on specific types of, 4-5. *See*
    also: Hazard(s)
hydraulic boom, 14
injury and death rates for, 2-3
latticework boom, 14
mobile hydraulic, 13
mobile rough-terrain, 54-57
monorail and underhung, 17
number of, in use, 2
overhead bridge, 125-27
pedestal hydraulic boom, 15
principles for reducing hazards on,
    7-9. *See also:* Hazard(s)
straddle, 18
track-mounted, 16-17, 81-82
trolley boom, 15
truck-mounted, 14, 74
wheel-mounted telescoping boom, 13
Cribbing. *See:* Blocking
Cow catchers. *See:* Guarding *and* Track-
  mounted cranes

D

Danger zone
    as related to pinch points, 63. *See also:*
      Pinch points
    mapping the, 23-26, 30-31
    safety planning for, 8, 23-26
Daniels, Saltz, Mongeluzzi & Barrett, Ltd.,
  32
Derrick cranes, 19
    used at oil wells, 131-136
Design, 1-2, 7
    for safe access in overhead cranes, 61
      See also: Overhead bridge cranes
    modifying, to reduce the hazard, 2
      See also: Hazard(s) *and* System safety
    of a hoist that eliminates free fall, 98
      See also: Free fall
    of safe monorail stops, 129. *See also:*
      Stops
    to eliminate or minimize the
      hazard, 7. *See also:* Hazards

to eliminate pinch points, 64-65. *See also:* Pinch points
to withstand crushing of cab, 53. *See also:* Cab, crush resistant
W2 design criteria, 54. *See also:* Rollover protection systems
Dickie, D. E., 41
Double-ended pins. *See:* Boom disassembly

E

Easements, 26
Electric company. *See:* Utility companies
Employee training. *See:* Operator competency *and* Training
Emergency exit systems, 57-58
quick release windows for, 57-58

F

Fail-safe systems, 8
Fatalities, workplace, 2
Flatbed-truck mounted cranes. *See:* Truck-mounted cranes
Flying or swinging objects. *See:* Cab(s) *and* Guarding
Footings, 122-23, 131-32. *See also:* Upset
Free fall, 97-98. *See also:* Hooks *and* Load loss
Free-wheeling. *See:* Controls

G

Guarding
of all moving or rotating parts (covers), 69
of controls, 76-78
of electrical conductors, 125
of exhaust pipes, 59
of sheaves, 88
of windows with screens and grills, 58
pinch points, 65
with wheel bumpers (cow catchers) 81, 83
Guyline anchor systems. *See:* Anchor systems
Guywires. *See:* Anchor systems

H

Hammerhead cranes. *See:* Cranes, fixed
Handholds, 56, 59
Hazard(s). *See also:* Design *and* System safety
active, 7
armed, 7
associated with most cranes, 4, 3-4
caused by substitution of parts, 140
cost of uncontrolled, 10, 32. *See also:* Litigation
detection systems for, 8. *See also:* Warning systems
dormant, 8
excuses for not applying prevention measures to reduce, 2-3
free flow of information about, 4. *See also:* Litigation
general prevention procedures for, 147-148
guarding the, 7. *See also:* Guarding
inherent, 2
insulating the, 8
litigation as a valuable source of information about, 2, 107. *See also:* Litigation
on specific types of cranes, 3-4
on-site job analysis for, 148
prevention through system safety analysis, 7-11. *See also:* System safety
High lifts. *See:* Blind lifts
Hoist blocks. *See:* Load loss
Hoist cable. *See:* Cables, binding in boom hoist
Hoists. *See:* Monorails and underhung cranes
Hooks, 91-99. *See also:* Load loss
"killer", 91, 96
positive-type safety latches for, 91, 96
two-blocking and, 91. *See also:* Two-blocking
Human factors. *See:* Human performance
Human performance. *See also:* Controls *and* System safety
issues in control design, 98
limitation of, in estimating visually safe clearances, 23
in use of remote controls, 74
range of normal, 23
when an operator must use two controls, 47
Hydraulic boom cranes. *See:* Mobile hydraulic cranes

I

Inspections, need for annual, 151
Insulated links, 23
Interlocks, 8, 41-45
Investigation of injury, 140

J

Jib booms
>a system-safety analysis of, 102
>attachment devices for stowing, 102
>disconnecting, for storage, 102

K

Kaplan, Martin N., 23
Kinking. *See:* Cables

L

Latticework boom cranes. *See also:* Boom
>disassembly *and* Boom failure *and* Cables
>*and* Transportation of cranes
>>disassembly of, 106-110
>>prone to two-blocking, 47, 49. *See also:*
>>Two-blocking
Lawsuits. *See:* Litigation
Licensing, of crane operators. *See:* Operator
>competency
Lift cranes. *See:* Straddle cranes
Lifting
>capabilities, 148
>variables affecting capacity for, 35, 37
>when to use tower crane for, 101
Lifts, *See:* Blind lifts
Litigation, 24
>as a valuable source of hazard
>>information, 2, 24
>initial attempt to conceal hazard data
>>during, 32
>restricting the transfer of important
>>injury data, 107
LMI (Load Moment Indicator). *See:* Load-
>sensing devices
LML (Load Moment Limiter). *See:* Load-
>sensing devices
Load charts, 37
Load loss, 91-99. *See also:* Free-fall *and* Hooks
>cause of, on straddle cranes, 99
>hoist blocks and, 91
Load-measuring systems, 37-43. *See also:*
>Load-sensing devices *and* Rated capacity
>>American requirements vs. Western
>>>European, 37
>>retrofit programs for, 39
>>role of, in preventing turntable
>>>undocking, 119
>>standards for, 37-43
>>U.S. Military requirements for, 38

Load-moment devices. *See:* Load-sensing
>devices
Load-sensing devices. *See also:* Load-
>measuring systems
>>as standard equipment to avoid
>>>overloading, 119
>>required in the Netherlands, 38
Loading cranes for transport. *See:*
>Transportation of cranes
Loads, unknown, 148
Lockout systems, for overhead bridge cranes,
>126-127
Longshoring Standards
>1918.63, 86
>1918.74(a)(2), 41
>1918.74(a)(7), 105
>1918.74(a)(9), 37
>1918.74(a)(10), 64
>1918.74(c), 64
Lowrance, William W., 8

M

Maintenance, 139-141
>need for preventive, 151
>record-keeping systems for, 139
Manuals, operator's. *See:* Operator's manuals
Manufacturers' responsibilities
>to alert crane owners and users of
>>possible accelerated wear and
>>damage to the cable, 86
>to conduct service tests on hoist line
>>slack, 87
>to determine why critical components
>>are continually needing
>>replacement, 139
>to equip cranes with anti-two-blocking
>>systems, 48
>to fulfill their span of safety control,
>>10
>to keep crane owners informed of the
>>injury experience of their cranes,
>>140-41
>to maintain ongoing record of
>>renovation and improvements, 139
>to make a system safety analysis of jib
>>boom storage attachment devices,
>>102
>to notify owners of new parts and
>>components, 140
>to promote sales of user-friendly load-
>>measuring systems, 37, 41
>to provide interlocks, 41-42

to reexamine sheave sizes on future designs, 86

to use lifting hooks with safety latches, 96

to warn users of hazards, 9-10. *See also:* Operator's manual

when a known safeguard would overcome a hazard, 3

Marine Terminal Standards
1917.42 (b)(2), 87
1917.42(b), 86
1917.45, 58, 92, 104, 107
1917.45(d), 105
1917.45(e)(1), 70
1917.45(f)(1), 72
1917.45(f)(7), 41
1917.45(f)(9), 125
1917.45(i)(2), 64
1917.45(i)(5), 24
1917.46, 37

MIL Standards
MIL-C-23877B (YD), "Cranes, Hydraulic, Wheel-Mounted, 5 & 12½ Ton Capacities," 50
MIL-C-27718B (USAF), "Crane, Truck Mounted A/S32A-16," 38, 49
MIL-C-27718C (USAF), "Crane, Hydraulic, Truck Mounted, A/S32A16, 16,500 Pound Crane Capacity, 6x4, GED," 38, 49
MIL-C-27840B (USAF), "Crane, Truck Mounted A/S32H-11," 38
MIL-C-27840C (USAF), "Crane, Truck Mounted A/S32H-11, 4,000 Crane Capacity, 4x4," 38, 49
MIL-C-28564(YD), "Cranes, Hydraulic, Truck Mounted, 8x4, 25-, 40-, and 55-ton Capacities," 38
MIL-C-28614(YD), "Crane, Hydraulic, Rough-Terrain, 25-ton Capacity, Full Revolving, Wheel-Mounted, 4x4, DED," 38
MIL-C-28614A(YD), "Cranes, Hydraulic, Full-Revolving, Wheel Mounted, 4x4, DED," 38, 50
MIL-C-28616(YD), "Crane, Shovel, Convertible, Crawler-Mounted, Commercial DED," 49

MIL-C-28622(YD), "Crane, Hydraulic, Wheel Mounted, 4x4, Full Revolving, 45-ton Capacity, DED," 38
MIL-C-28629(YD), "Crane, Hydraulic, Wheel Mounted, 4x4, Full Revolving, (Shipboard Aircraft Crash Salvage)," 38
MIL-C-29352(MC), "Crane, Wheel-Mounted: Hydraulic, Rough-Terrain, 30-ton Capacity, Full Revolving, 4x4, DED," 38
MIL-C-52326B, "Crane, Crawler Mounted: 60-ton at 25-foot Radius, Diesel-Engine-Driven," 38
MIL-C-52341E(ME), "Crane, Wheel Mounted, Diesel Engine-Driven, 20-ton, Rough-Terrain," 38
MIL-C-529911(ME), "Crane, Truck Mounted, Hydraulic, 25-ton, Commercial Construction Equipment, (CCE)," 39
MIL-C-62135A(AT), "Crane: Material Handling," 39
MIL-C-62253(AT), "Crane: Material Handling (Rebuild)," 39
MIL-HDBK-759, 72, 98
MIL-STD-1472, 72, 97
MIL-STD-1472B, 72, 98
MIL-STD-803A-1(USAF), 72, 97
MIL-T-62089(AT), "Truck, Maintenance with Rotating Hydraulic Derrick, Air Transportable, 34,500 pounds, GVW, 6x4," 38, 39, 41

Military Specifications, 49

Mobile cranes, during transport, 111-114. *See also:* Transportation of cranes

Mobile hydraulic cranes, 13
limited vision during backing of, 80
prone to two-blocking, 47
stability of, on side lifts, 41
typical cab control station on, 73

Mobile and locomotive cranes, ANSI/ASME B30.5-1994, 24, 37, 49

Mobile rough-terrain cranes
stability index for, 40, 55
two-wheel/four-wheel steering on, 51. *See also:* Steering, two-wheel/four-wheel
upset criteria vs. crawler tractors, 54. *See also:* Rollover protection systems

Mongeluzzi, Robert J., 32

Monitoring systems, electrical and
mechanical, 8
Monorail and underhung cranes, inadequate
stops on, 129

N

National Electrical Safety Code, 22-24
National Commission for the Certification of
Crane Operators, 10, 156
Nip points, 69-70. *See also:* Pinch points

O

Operator competency, 143-45. *See also:*
Personnel competency *and* Training
certification for, 143
in-house licensing to assure, 143-44
role of, in reducing deaths and
injuries, 144
standards for, 144-45
Operator's manual
ability to withstand wind loading
discussed in, 137
boom buckling discussed in, 102
how to operate the emergency escape
system stated in, 57
procedures for crane transport
discussed in, 111
procedures for raising and lowering
tower cranes discussed in, 122
routine checks of the turntable
assembly required by, 119
safe work procedures for straddle
cranes included in, 83
stability index stated in, 55
warning users of hazards in, 9. *See
also:* Manufacturers' responsibilities
written lockout procedure discussed
in, 126
Operator's station, 53-58. *See also:* Cab *and*
Vision
OSHA requirements
for ground fault interrupters, 23-24
general, for cranes and derricks, 5
OSHA Standard
1910.132, 59
1910.145, "Specifications for Accident
Prevention Signs and Tags," 9
1910.147, 126
1910.179(b)(4), 137
1910.179(b)(5), 37
1910.179(b)(6)(ii), 81
1910.179(e)(4), 82

1910.179(e)(5), 87
1910.179(e)(6)(i), 69
1910.179(e)(6)(ii), 69
1910.179(e)(6)(iii), 69
1910.179(g)(1), 125
1910.179(g)(2), 125
1910.179(h)(1)(ii), 87
1910.179(h)(1)(iii), 87
1910.179(h)(4)(i), 82
1910.179(i), 79
1910.179(j)(2)(iii), 92
1910.179(j), 129
1910.179(l)(2), 126
1910.179(l)(3)(iii)(a), 92
1910.179(m)(1), 86, 87
1910.179(n)(3)(iv), 105
1910.179(n)(3)(vi), 107
1910.179(n)(3)(x), 107
1910.180(a)(16), 104
1910.180(c), 37
1910.180(d)(3)(v), 92
1910.180(g)(1), 86, 87
1910.180(h)(3)(iv), 105
1910.180(h)(3)(vi), 107
1910.180(h)(3)(vii), 37, 120
1910.180(h)(3)(ix), 41
1910.180(h)(3)(xiv), 81
1910.180(h)(3)(xv), 104
1910.180(h)(4), 107
1910.180(i)(6), 64
1910.180(j), "Operating Near Electric
Powerlines," 22
1910.181(c), 37
1910.181(j)(2)(ii), 92
1926.1000, 53
1926.1003, 53
1926.16, "Rules of Construction," 79
1926.181(g)(1) 86, 87
1926.20(a) and (b), Subpart C,
"General Safety and Health
Provisions," 72, 101, 116, 121
1926.20(b), 101
1926.200, Subpart G, "Signs, Signals,
and Barricades," 9
1926.201, 80
1926.550, 48, 59
1926.550, Subpart N, 48
1926.550(a), General Requirements,
107
1926.550(a)(1), 37, 101-102
1926.550(a)(1), Subpart N, 88
1926.550(a)(6) 86, 87, 141
1926.550(a)(7), 86, 87
1926.550(a)(9), 64

1926.550(a)(15), 22, 24, 33
1926.550(a)(15)(v), Code of Federal
    Regulations, Subpart N, 23
1926.550(a)(17), 104
1926.550(d)(3), 80
1926.550(g)(4)(iv)(B), 92
1926.600(a)(3)(i), 107
1926.601(b)(4), 81
1926.955(e)(5), 22-23
No. 3067, 70
Outriggers. *See also:* Footings *and* Upset
    blocking and cribbing for, 41
    collapse of, 41
    extension of, 45
    failure to use, 40
    inadequate floats or pads on, 42
    lack of, on self-propelled aerial lifts,
      116
Overhead bridge cranes
    absence of lockout systems on,
      126-27
    exposed electrical trolleys on, 125-26
Overloading. *See:* Upset

P

Panic bars, 81-82
Pedestal hydraulic boom cranes, 74
Peening. *See:* Cables
Personal protective equipment, 9
Personnel competency, 143-45
    for riggers and signalers, 145
    requirements to ensure, 143-45
Pick-and-carry operations
    limited vision during, 80
    locating routes for safe, 22
    need for additional safeguards for, 23
    powerline contact during, 22, 23
    use of proximity alarm in, 23-25
    use of signalers during, 80-81
Pinch points, 63-70
    defined, 63
    designing to eliminate, 64-65
    in extendible counterweight
      systems, 115-16
    inherent in many mobile cranes, 63
    location of primary, 63
    patents for preventative devices, 65
Pins. *See:* Boom disassembly
Planning
    for blind lifts, 79. *See:* Blind lifts
    lack of, in crane use, 2
    pre-construction, 147

Powerline contact, 21-34
    by aerial lifts, 32-33
    by hoist or boom, 21
    by pedestal hydraulic boom cranes,
      74
    electrical remote-control systems
      (control boxes) and, 24
    fatality data on, 25
    injury data on, 21
    litigation based on, 25
    pre-job safety planning to avoid, 21
    re-closure systems and, 21
    types of cranes involved in, 26
    voltages for, 26
Proximity alarms, 24-26, 74. *See also:* Back-up
    safety devices *and* Warning systems

R

Radio-controlled controls. *See:* Remote
    controls
Radios, two-way. *See:* Communications
Rated capacity. *See also:* Load-measuring
    systems
    converting to a higher, 139-140
    lift in excess of, 119
    prevent the exceeding of, 37
Rating chart. *See:* Load charts
Redundancy, 8, 23
Remote controls, 99
    hand-held, on wire tethers, 74
    non-conductive, 24
Removable counterweights, 115
Restraint system, 8, 54, 112
Rigging practices, 148
Roebling handbook, 85
Rollover protection systems (ROPS)
    equipping mobile, rough-terrain
      cranes with, 54
    standards, 53
    substantially reduce deaths and
      injuries, 53
    U.S. Army Corp of Engineers criteria
      for, 54
    W2 design criteria, 54
Rope, wire. *See:* Cables
ROPS. *See:* Rollover protection systems
Rough-terrain cranes. *See:* Mobile rough-
    terrain cranes

S

Society of Automotive Engineers (SAE) Standard
- J115, "Safety Signs," 9
- J159, "Crane Load Moment System," 39
- J185, "Access Systems for Off-road Machines," 57, 60
- J220, Crane Boomstop (March, 1991) 104
- J375, "Radius-of-Load and Boom Angle Measuring System," 39
- J376, "Load Indicating Devices in Lifting Crane Service," 39
- J673, "Automotive Safety Glazing," 58
- J674, "Safety Glazing Materials-Motor Vehicles," 58
- J881, "Lifting Crane Sheave and Drum Sizes," 86, 89
- J959 (October, 1980), "Lifting Crane, Wire-Rope Strength Factors," 87
- J983, "Crane and Cable Excavator Basic Operating Control Operations," 72
- J983, "Crane-Shovel Basic Operating Control Arrangements," 98
- J983, 1971 SAE Handbook, SAE Recommended Practice, 72
- J987 (October, 1980), "Crane Structures — Method of Test," 101
- J999, Crane Boom Hoist Disengaging Device (February, 1985), 104
- J1063 (October, 1980), "Cantilevered Boom Crane Structures — Method of Test," 101
- J1180, "Telescoping Boom Length Indicating System," 39
- J1289, "Mobile Crane Stability Ratings," 39
- J1305, "Two-Block Warning and Limit Systems in Lifting Crane Service," 49

Safety and Health Requirements Manual, EM 385-1-1, 24, 37, 40, 50, 61, 64, 76, 80, 81, 82, 83, 87, 90, 104, 105, 107, 122, 123, 126, 137, 144, 145

Safety systems. *See:* Hazards, prevention concepts *and* System safety

Scardino, A. J. 131

Screens. *See:* Guarding

Seat belts. *See:* Restraint system

Sensors, 8

Sheaves, 85-90. *See also:* Cable(s)
- wear of plastic vs. metal, 88

Side pull, 105-106
- causes of, 105
- sensing systems, 105

Sigma Associates, 131

Signals. *See also:* Blind lifts *and* Communications
- appropriate hand, 148-151
- for blind lifts, 79

Site plans. *See:* Danger zone, mapping

Slack. *See:* Cables

Slope indicators. *See:* Warning systems

Sluing, 37, 42, 101

Specialized Carriers and Riggers Association, 144

Stability index, 40, 55. *See also:* Mobile rough-terrain cranes

Steering, two-wheel/four-wheel, 55

Stops. *See:* Monorail and underhung cranes

Straddle cranes
- load loss on, 99. *See also:* Load loss
- site preparation for, 99
- use of travel alarms on, 83
- vision compromise on, 83

System safety. *See also:* Hazard(s)
- as an approach to hazard prevention, 7, 64
- human performance as a principle of, 7, 9-11. *See also:* Human performance
- outline of, principles, 9
- use of, in hazard analysis, 102, 104

T

Telephones. *See:* Communications, two-way radio use

Telescoping-boom cranes. *See:* Wheel-mounted telescoping-boom cranes

Television, closed circuit. *See:* Communications

Tipping load. *See:* Upset

Tower cranes, 121-23
- allowed to rotate freely in wind, 137
- checklist for the climbing of, 121
- drawings of structure supporting, 119
- footing failure in, 122-23
- most hazardous operating cycle in, 121
- upset of self-raising, 121-23

use of, to avoid boom failure, 101
Track-mounted cranes, 81-82
Training, 9, 99, 106. *See also:* Operator
competency
crane transport, 111. *See also:*
Transportation of cranes
employees to use blocking, 106
for handling of counterweights, 115
for operator competency reducing
deaths and injuries, 144
instructors, 145
to overcome inherent hazards, 74
Transportation of cranes, 111-14. *See also:*
Upset, soil failure
adjusting A-frames during, 112
bouncing during, 111
operator's manuals should contain
information on safe, 113
trailers and trucks susceptible to
sinking or being stuck during, 113
Travel alarms, 81-82. *See also:* Warning
systems
Travel mode of mobile crane. *See:*
Transportation of cranes
Truck-mounted cranes
load limits for, 36
upset of, 41
Turntable undocking, 119-21
inspection of turntable assembly to
prevent, 119
Two-blocking, 47-52. *See also:* Anti-two
blocking devices
causing latticework boom
failure, 47-48
death and injuries resulting from, 47
definition of, 47
failure of lifting hook after, 91
injury data, 52
types of cranes prone to, 47

U

Underhung cranes. *See:* Monorail and
underhung cranes
Undocking. *See:* Turntable undocking
United Underwriters Laboratories, 8
Unloading cranes. *See:* Transportation of
cranes
Upset, 35-45. *See also:* Outriggers *and*
Turntable undocking
aerial lift or elevated work
platform, 44

aerial lift, when outriggers not
extended, 44-45
audible slope indicators to warn
of, 116
due to toppling of latticework
boom, 103
failure to use outriggers causing, 40-42
of counterweight-stabilized aerial
lifts, 116
overloading as trigger for turntable
undocking and, 119
soil failure, 40-42, 117, 131. *See also:*
Footings; Soil stability *and*
Transportation of cranes
statistical data, 36, 41
structural failure causing, 40-42
tipping loads and, 36
types of cranes prone to, 54
Utility companies
absence of warnings in easements
from, 25-26
buying criteria for lifts for, 33
de-energizing powerlines by, 74
information on mapping the danger
zone provided by, 25
lockout of automatic re-closures by,
21
placement of powerlines by, 22

V

Vibratory pile drivers, 37
Vision, 79-84
blind spots, 79
compromise on large track-mounted
cranes, 81-82
compromise on overhead bridge
cranes, 126
compromise on straddle cranes, 83
in pick-and-carry mode, 80
lack of, in crane backing, 80-81

W

Walkways, unsafe, 61-62
on overhead bridge cranes, 61
Warning devices. *See:* Warning systems
Warning signs and labels, 9, 57
for boom disassembly, 106, 109-110
for the safe handling of
counterweights, 116
to warn users of hazards, 9
Warning systems 8, 79

audible slope indicator, 116
back-up alarm, 81
for side pull, 105
proximity alarm, 23, 25, 74
travel alarm, 79, 83
Wheel bumpers, 83. *See also:* Guarding
Wheel-mounted telescoping-boom cranes,
    14, 48-49
equipped with anti-two-blocking
    devices, 48-49
Wind
allowing tower crane to rotate freely
    in, 137
and other weather factors, 137
loading, 131, 137
Wire rope. *See:* Cables
Worksite, as danger zone. *See:* Danger zone
Wylie Weighload, 40

About the Author

During his over fifty-year career, David V. MacCollum has had the opportunity to examine or review over 10,000 crane-related accidents that have caused injury or death. His approach to crane safety is to identify both hazardous circumstances and design shortcomings. As a result of his research on tractor rollover for the Corps of Engineers in 1958, the Corps adopted his design recommendations for ROPS ten years before industry standards were published. ROPS now saves several hundred lives a year. His twenty-year advocacy for load-moment devices and anti-two-block systems on cranes has led, in some measure, to universal acceptance. His views on making cranes safer in the future are reflected throughout this book.

His experience with cranes began in World War II as a Navy crane operator in the Philippines and the Marshall Islands. After the war, as a student at Oregon State College, he worked in the summers in logging camps and operated a variety of log loaders. After obtaining a Bachelor of Science degree, he began his career as a safety engineer with the Accident Prevention Division of the State of Oregon's Industrial Accident Commission. In this position he gathered information about hazard control and crane safety for employers in high-risk industries, including construction, logging and mining.

In 1955, as a safety engineer for the U.S. Department of Army Corps of Engineers, he developed field-testing criteria to ensure safe crane activity on construction projects. In 1963, he was appointed Safety Director for the U.S. Army Test and Evaluation Command at Fort Huachuca, Arizona, and was responsible for system safety analysis of newly developed army equipment.

In 1969 he became Safety Director for the U.S. Army Strategic Communications Command with sixteen subordinate commands worldwide. In this post he became familiar with European and Pacific Rim crane safety requirements. The same year he was appointed to the U.S. Secretary of Labor's Construction Safety Committee under the Construction Safety Act — the forerunner of OSHA — and helped draft the original *Subpart N, Cranes, Derricks, Hoists, Elevators and Conveyors*, 1926.550, now part of OSHA.

In 1972 he established a safety research consulting practice, initially assisting colleges and universities in developing safety engineering and continuing-education programs. As a consultant to a Swiss-German firm that designed tunnel and underground support systems, he became increasingly aware of European design safety standards for cranes. In 1975-76 he served as president of ASSE.

In 1999 he became a Fellow in the American Society of Safety Engineers, an honor bestowed upon him for his achievement and service in the safety profession.

After more than 54 years as an active advocate for engineering solutions, he still believes that safe design is one of the best solutions for overcoming hazardous conditions. This belief has become the focal point of his thinking after more than 50 years devoted to conducting investigations, making analyses, researching available engineering improvements and giving testimony in injury litigation concerning cranes and other hazardous construction and industrial equipment.

In this new millennium he has been instrumental in founding the Hazard Information Foundation, Inc. (HIFI) to preserve his and other engineers' extensive libraries of reference material pertaining to the prevention of hazards through safe design of equipment and facilities and use of safeguards to protect workers.

His ongoing goal is to focus the talents and expertise of engineers in using their knowledge in the physical sciences to develop safer design and accessories and promote new construction methods to overcome the inherent hazards that have plagued construction, mining, and other high-hazard activities for years.

Notes

Notes

Notes

Notes

Notes

Notes